STUDY GUIDE

for the Third Edition of

Keeton's Biological Science

STUDY GUIDE

for the Third Edition of

Keeton's Biological Science

CAROL HARDY MC FADDEN

W · W · NORTON & COMPANY · NEW YORK · LONDON

Copyright © 1980, 1979, 1978 by W. W. Norton & Company, Inc.

Printed in the United States of America

All Rights Reserved

First Edition

ISBN 0 393 95028 X

2 3 4 5 6 7 8 9 0

CONTENTS

To the Student

Chapter 1	Introduction	1
Chapter 2	Some Simple Chemistry	7
Chapter 3	Cells: Units of Structure and Function	17
Chapter 4	Energy Transformations	27
Chapter 5	Nutrient Procurement and Processing	35
Chapter 6	Gas Exchange	43
Chapter 7	Internal Transport	49
Chapter 8	Regulation of Body Fluids	59
Chapter 9	Chemical Control	67
Chapter 10	Nervous Control	81
Chapter 11	Effectors	93
Chapter 12	Animal Behavior	99
Chapter 13	Cellular Reproduction	111
Chapter 14	Patterns of Inheritance	117
Chapter 15	The Nature of the Gene and Its Action	131
Chapter 16	Development: Aspects of Cellular Control	139
Chapter 17	Development: From Egg to Organism	145
Chapter 18	Evolution	153
Chapter 19	Ecology	165
Chapter 20	The Origin and Early Evolution of Life	179
Chapter 21	Viruses and Monera	185
Chapter 22	The Protistan Kingdom	191
Chapter 23	The Plant Kingdom	197
Chapter 24	The Fungal Kingdom	205
Chapter 25	The Animal Kingdom	209
Answers		221

TO THE STUDENT

Education research has shown that almost 90 percent of the material a student hears or reads is forgotten within a few months—a most discouraging fact! Fortunately, retention can be increased by working with the material and by explaining it to others. This *Study Guide* gives you the opportunity to do just that; it requires you to *use* the material you are studying. To answer many of the questions you must synthesize, analyze, and interpret information. The *Study Guide* is designed to help you focus on the important concepts you need to learn the material presented in *Biological Science*.

Each chapter of the *Guide* has five parts: Key Concepts, Objectives, Summary, Questions, and Answers.

Key Concepts Each chapter begins with a list of its key concepts, which provide a framework for facts that the chapter presents.

Objectives Each chapter provides clear objectives. These objectives describe what you should be able to do when you have mastered the material. They are not all-inclusive; they are directed toward the chapter's major concepts. They suggest a plan for study, define the goals toward which you should aim, and help you organize the content of the chapter.

When you begin studying, first read over the key concepts and the objectives. Then begin reading the chapter in your text. Keep the *Study Guide* handy while you are reading and try to keep the key concepts in mind. At the end of each section, stop and try to relate what you have read to the key concept. Look again at the objectives. Have you met them? Reread the sections that are unclear.

Summary The *Study Guide* offers a summary of each chapter in the text. Since many of the chapters are long, the summaries can help you master the material.

Questions This section is divided into three parts: "Testing recall," "Testing knowledge and understanding," and "For further thought." The "Recall" questions require you to remember the basic facts and vocabulary presented in the chapter. The questions are varied: true and false, matching, fill-in-the-blank, and multiple choice. When you have finished reading the chapter, do this section and check your answers. These questions will help you identify areas that need further study. Once you answer the "Recall" questions correctly, you can go on to the next section.

In general, the questions in "Testing knowledge and understanding" are more demanding. You will need to recall information and also analyze and integrate it. The questions in this section are multiple choice. You should choose the *best*, most specific answer from among the choices given. When you choose the answer you think is correct be sure you understand why each one of the other possibilities is wrong. All these questions have been tested and evaluated; they are drawn from ten years of examinations in the introductory biology course at Cornell University.

The "For further thought" section also requires you to apply your knowledge. To answer these essay-type questions you must be able to recall the appropriate information and apply it to the situation. These questions ask you to use your knowledge effectively.

Answers Answers for the "Recall" and "Knowledge" questions are provided on pages 221–232. Page references in the text are given for each question, so you can look up the material you have missed.

Acknowledgements I am indebted to Kraig Adler and Neil Campbell for many of the questions used in the *Study Guide* and to Verne E. Rockcastle for his help in developing the format. I also acknowledge the comments, suggestions, and criticisms of the introductory biology students at Cornell in the 1978 summer session and the 1978–1979 academic year.

I am especially grateful to William T. Keeton for giving me the opportunity to develop the *Study Guide* and for all of his advice during the preparation.

Finally, I thank my mother for her careful reading of the manuscript, and my children, Daniel and Jean, who have had to live with a very busy and preoccupied mother for the past two years.

I hope the *Study Guide* will help you in your study of biology. Your comments and suggestions would be most welcome.

Ithaca, New York
December, 1979   Carol Hardy McFadden

Chapter 1

INTRODUCTION

KEY CONCEPTS

1 All science is concerned with the material universe, seeking to discover facts about it and to fit these facts into theories or laws that will clarify the relationships among them.

2 Organisms change with time; the course of this evolutionary change is determined by natural selection.

OBJECTIVES

After studying this chapter and reflecting on it, you should be able to

1 Discuss the scientific method and its applications and limitations.

2 Explain why the physical sciences entered their modern era almost two hundred years before biological science entered its new era.

3 Explain why the year 1859 is taken as the beginning of the new era in biological science.

4 Discuss the two parts of Darwin's theory of evolution and give the five basic assumptions upon which it rests.

5 Name the five kingdoms of organisms and give the distinguishing characteristics of each.

6 Give one important characteristic of each of the following groups of plants: green algae, brown algae, red algae, mosses, and vascular plants.

7 Give one important characteristic and one example of each of the following groups of animals: coelenterates, flatworms, molluscs, annelids, arthropods, echinoderms, and chordates.

SUMMARY

Biological science is the study of life. All science is concerned with the material universe, seeking to discover facts about it and to fit these facts into theories or laws that will clarify their relationships.

Scientific method Science begins with *observations* of objects or events. A scientist gets all information through his sense organs. Once a scientist has formulated the question he wants to ask, and has made repeated, careful observations in an attempt to answer it, he uses these observations to formulate a generalization. The generalization must then be tested, and as testing proceeds, the generalization must be altered or even abandoned when necessary to conform with the evidence. Eventually, if all evidence continues to support the new idea, it may become widely accepted as probably true and so be called a *theory*. However, the testing never completely stops; no theory in science is ever absolutely and finally true.

The insistence on testability in science imposes limitations on what it can do. Science cannot make value and moral judgments.

The rise of modern biological science The physical sciences underwent explosive growth during the sixteenth and seventeenth centuries, but it was not until 1859 that biological science entered its modern era. In that year, Charles Darwin published *The Origin of Species*, in which he proposed his theory of evolution by natural selection. To this day the theory remains one of the most important unifying principles in all biology.

Darwin's theory Darwin postulated that all organisms living today have descended by slow, gradual changes from ancient ancestors quite unlike themselves. The course of this evolutionary change is determined by natural selection. Darwin's theory depends upon five basic assumptions:

1 Many more individuals are born in each generation than will survive and reproduce.
2 There is variation among individuals.
3 Individuals with certain characteristics have a better chance of surviving and reproducing than individuals with other characteristics.
4 At least some of the characteristics resulting in differential reproduction are heritable.
5 Long spans of time are available for slow, gradual change.

The variety of organisms According to the fossil record, the most primitive organisms known (the bacteria and blue-green algae) date back over 3 billion years. Innumerable different kinds of organisms have evolved from these. The process by which the different organisms have evolved begins generally with geographic separation of some members of an interbreeding population. The new population will inevitably be in a slightly different environment, and natural selection may favor other characteristics than those of the original population. Over time, the differences between the populations become such that they can no longer interbreed, even if the geographical barrier between them disappears. It is this process, repeated over billions of years, that has given rise to the various groups of organisms seen today. The various groups can be classified into five kingdoms: Monera, Protista, fungi, plants, and animals. Those organisms belonging to the Monera and Protista are unicellular or colonial; the others are all multicellular.

The kingdom *Monera* includes two groups of unicellular organisms: the bacteria and the blue-green algae. Members of the Monera differ from those of the four other kingdoms in that their cells lack the membrane-enclosed nucleus and other membranous structures present in the cells of other organisms.

The kingdom *Protista* includes a wide variety of unicellular or colonial organisms. Two major groups of Protista are those that process chlorophyll and are photosynthetic (the plantlike protists), and those lacking chlorophyll (the animal-like protists, or *Protozoa*).

The organisms belonging to the plant kingdom all have cells with rigid cell walls and subcellular structures containing chlorophyll. Thus they are photosynthetic and can synthesize their own food. The red, brown, and green *algae* are aquatic plant groups that show little tissue differentiation. The land-plant groups probably evolved from ancestral flagellated green algae.

The *mosses* are not completely free of the aquatic environment and must live in moist habitats. The *vascular plants* (often called the higher land plants) show the greatest tissue specialization; in particular, they possess vascular tissue, which provides conduits for transport of materials from one part of the plant to another, making these plants less dependent on water in the surrounding environment. Ferns, conifers (*gymnosperms*), and flowering plants (*angiosperms*) are important vascular plants. The ferns are the most primitive of these groups, the angiosperms the most advanced. The angiosperms are customarily divided into two subgroups: the monocotyledons (grasses and grasslike plants) and the dicotyledons.

Like plants, *fungi* have cell walls. However, they lack chlorophyll and cannot manufacture their own food; they must obtain their nutrients already synthesized. Fungi cannot ingest particulate food; they depend entirely on absorption of nutrient molecules.

Animal cells differ from plant cells in the lack of rigid cell walls and in their mode of nutrition. Animals ingest particulate food. Seven of the animal groups will be frequently referred to throughout the book; you should become familiar with them.

The *coelenterates* are radially symmetrical animals whose bodies are composed of two distinct tissue layers. Their digestive cavity has only one opening. The *flatworms* have a similar digestive cavity but are bilaterally symmetrical and composed of three distinct tissue layers. The *molluscs* are fairly complex organisms, most of which possess shells. The *annelids* are the segmented worms. All of the *arthropods* have jointed legs and a hard outer skeleton; they are an immense group of very advanced animals.

The *echinoderms* and *chordates* are two closely related groups. The echinoderms are radially symmetrical marine animals whose members include the sea stars, sand dollars, and sea urchins. Chordates include the *vertebrates*, which comprise all animals possessing a hard internal skeleton: the fish, amphibians, reptiles, birds, and mammals.

A helpful way to remember the various animal groups is to arrange them in an order based on their supposed evolutionary relationships, with the most primitive (like the ancestral condition) groups at the bottom, and the most advanced (unlike the ancestral condition) groups at the top:

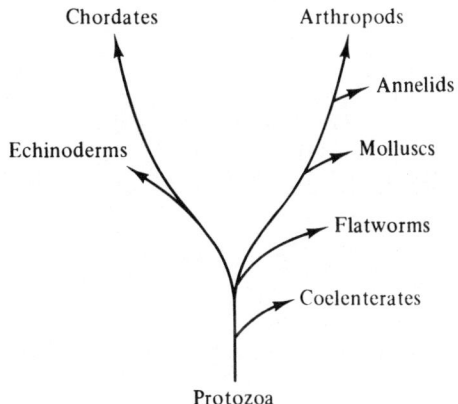

Those organisms at the bottom of the diagram are sometimes called the lower animals; those at the top, the higher animals.

The evolutionary relationships among the vertebrates may be diagrammed as follows:

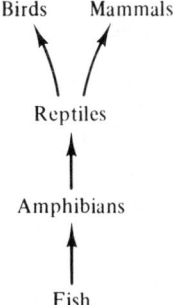

QUESTIONS

Below are listed the five kingdoms of organisms used in the classification system presented in your textbook. Answer the following questions with the appropriate letter(s); more than one kingdom may answer each question.

a Animalia   d Plantae
b Fungi   e Protista
c Monera

1 Which kingdoms have only unicellular or colonial organisms?

2 Which kingdoms have multicellular organisms?

3 Which kingdoms include organisms whose cells have cell walls?

4 Which kingdoms include at least some organisms that can synthesize their own food?

5 Which kingdoms include organisms that ingest particulate food?

6 In which kingdom are nearly all the organisms photosynthetic?

7 In which kingdom(s) are all the members totally dependent on absorption to obtain their high-energy nutrients?

Below are listed some of the important animal groups that you are expected to become familiar with. Match each of the characteristics with the appropriate group or groups of animals.

a annelids   e echinoderms
b arthropods   f flatworms
c chordates   g molluscs
d coelenterates

8 digestive cavity with one opening

9 two distinct tissue layers

10 jointed legs and hard outer covering

11 radial symmetry

12 shells

13 internal hard skeleton

14 segmented worms

Referring to the list above, match each animal with the group it belongs to.

15 grasshopper
16 clam
17 lobster
18 planaria
19 hydra
20 starfish
21 fish
22 crayfish
23 jellyfish
24 snake
25 frog
26 human being
27 earthworm

Use the following key to answer questions 28–36.

a algae   c mosses
b fungi   d vascular plants

28 Which group are the dominant land plants?

29 Which group obtains its nutrients only by absorption?

30 Which group is divided into angiosperms and gymnosperms?

31 Which group are aquatic plants that show little tissue differentiation?

32 Which group are land plants that show little tissue differentiation?

33 Which group shows the greatest tissue specialization?

34 Which group includes the flowering plants?

35 Which group lacks chlorophyll?

36 Which group possesses an effective transport system?

Choose the one best answer.

37 For consistency with the theory of natural selection as the agent of evolution, it is necessary to postulate that

  a in each generation, all individuals well adapted for their environment live longer than those not so well adapted.
  b the deaths of individuals occur completely at random with respect to the environment.
  c some of the reproductive success of individual organisms is heritable.
  d most deaths of individual organisms occur soon after birth.
  e more than half of the poorly adapted individuals must be eliminated by natural selection in each generation.

38 In formulating his theory of evolution, Charles Darwin collected and synthesized data on a variety of subjects. Which one of the following is *not* a line of evidence used by Darwin in formulating his theory?

  a the existence of fossils of extinct animals
  b the similarities of living organisms
  c creation of the earth in 4004 B.C.
  d gradual morphological changes in fossils over time
  e structural changes in domestic animals and plants

Chapter 2

SOME SIMPLE CHEMISTRY

KEY CONCEPTS

1. Both living and nonliving matter is made up of the same fundamental particles. The only difference between living and nonliving things seems to be in the way these basic materials are organized.

2. Living organisms are integral parts of the physical universe and must obey the fundamental laws of chemistry and physics.

3. Weak chemical bonds are important in stabilizing the shape of many large molecules and in holding them together.

4. Water molecules are polar and have a strong tendency to form hydrogen bonds with one another; the consequent ordering of the molecules gives water many of its special properties, making it the medium for life.

5. All of the complex organic molecules are composed of many simpler building-block molecules, bonded together by condensation reactions; they can be broken down to their building-block molecules by hydrolysis.

6. The amino acid content and sequence of a protein determine its three-dimensional shape; alteration in the shape of the protein can lead to a change in its biological function.

7. Enzymes are globular proteins that catalyze the chemical reactions within organisms; their catalytic activity depends on their three-dimensional shape.

OBJECTIVES

After studying this chapter and reflecting on it, you should be able to

1. Relate the structure of an atom to its chemical properties and the type of chemical bond it forms.

2. Distinguish between ionic and covalent bonds and the roles they play in the formation of biological compounds.

3. Explain why the most common elements in the bodies of living organisms (C, O, H, N) form highly stable molecules.

4. Explain the crucial role of weak chemical bonds (such as hydrogen bonds, van der Waals and hydrophobic interactions) in the organization of living materials.

5. Distinguish between polar and nonpolar molecules and relate the characteristics of each to their solubility in water.

6. Relate the polarity of the water molecules and their tendency to form hydrogen bonds to the special properties of water that make it the basis for life on earth.

7. Define acid, base, buffer, pH, and explain why the pH must be kept within certain limits within the cell or organism.

8. Identify examples of each of the four main classes of biologically important organic molecules and the building-block units of which they are composed. Give a major function of each of these classes.

9. Write a condensation and hydrolysis reaction, given the reactants.

10. List four factors that influence the rate of chemical reactions and explain how each of these affects the rate. Know which of these factors are important in governing the rate of reactions within the human body.

11. Explain why the structure and biological activity of a protein are determined by the sequence of the various amino acids in its polypeptide chains.

12. Distinguish between an exergonic and an endergonic reaction and explain why these reactions are usually coupled in living systems.

13. Explain why the three-dimensional structure of an enzyme is the key to its activity, and the role that temperature, pH, and inhibitors play in altering enzyme activity.

SUMMARY

The matter of the universe is composed of a limited number of basic substances called *elements*. In any element the smallest unit that cannot be subdivided by ordinary chemical means is the atom. *Atoms* are made up of a nucleus containing the positively charged *protons* and uncharged *neutrons*, surrounded by a cloud of negatively charged *electrons*. In a normal neutral atom the number of protons is equal to the number of electrons. The electrons are in constant motion outside the nucleus. The distance of an electron from the nucleus is a function of its energy; the higher the energy, the farther from the nucleus the electron is likely to be. The region in which the electron will probably be found is called the *orbital* of the electron. Only certain energy levels are possible for electrons in atoms. The lowest energy level can contain a maximum of two electrons, the next level eight. The chemical properties of elements are determined largely by the number of electrons at their outer energy level. Most atoms are in a particularly stable configuration when their outer energy level contains eight electrons. Atoms tend to form complete outer levels by reacting with other atoms.

Chemical bonds The attractive force that holds atoms together is a *chemical bond*. Only the electrons at the outermost energy level (the *valence* electrons) are ordinarily involved in the formation of chemical bonds. Chemical bonds may be *ionic*, in which case electrons are actually transferred from one atom to another, or they may be *covalent* bonds, in which electrons are shared between atoms.

In an ionic bond, a strong electron donor loses an electron to an electron acceptor. The element that has lost an electron now becomes a positively charged atom, or *ion*; the element accepting the extra electron becomes a negatively charged ion. The two ions are then attracted to each other by their opposite charges. The resulting ionic bond is strong in the solid state, but relatively weak in aqueous solutions.

Two particularly important ionic compounds are the *acids* and *bases*. An acid is a substance that increases the concentration of hydrogen ions (H⁺) in water, while a base decreases the hydrogen ion concentration. Acidity and alkalinity (basicity) are measured in terms of *pH*, a value reflecting the concentration of hydrogen ions. Living matter is extremely sensitive to pH; some or all of the function of living cells can be destroyed by even small changes in pH. *Buffers* are substances that tend to resist changes in pH by combining with or releasing hydrogen ions.

Covalent bonds are strong bonds; they can be broken apart only with relatively large amounts of energy; hence covalently bonded molecules are stable. Carbon, oxygen, hydrogen, and nitrogen are the most common elements in the bodies of living organisms, and these small atoms form particularly strong covalent bonds. Life, therefore, is based on elements that form highly stable molecules.

When a covalent bond is formed between atoms of the same element, the electrons are shared equally and the bond is said to be *nonpolar*. A *polar* bond is formed when atoms of two different elements share electrons. In this case the electrons are not shared equally, but tend to be pulled closer to one element than to the other, and this causes an asymmetry in the distribution of charge. In a water molecule the electrons are closer to the oxygen atom than to the two hydrogen atoms. Therefore the bonds are polar, with the oxygen side of the molecule more negative and the hydrogen ends of the molecule more positive. The whole water molecule is polar as a result of the polarity of the bonds within the molecule and the V-shaped arrangement of the atoms. The oxygen end of the molecule has two slight negative charges; it is electronegative. The two hydrogen ends each have a slight positive charge; they are electropositive.

In addition to ionic and covalent bonds, three other kinds of bond or interaction play an important role in biological systems. These are hydrogen bonds, van der Waals interactions, and hydrophobic interactions. Although these are weak bonds, they are important in stabilizing the shape of many of the large, complex molecules found in living matter.

Of particular importance is the *hydrogen bond*. Hydrogen bonds result when a hydrogen atom is shared by two electronegative atoms such as nitrogen or oxygen:

$$-C=O^- --\ ^+H-N-$$

The hydrogen is linked by a polar covalent bond to one of the electronegative atoms. Thus the hydrogen atom is electropositive and is attracted to an electronegative atom in another molecule.

Water Water molecules, because of their polarity, tend to form hydrogen bonds with each other. This attraction results in many of the special physical properties of water that have important implications for life processes. The capacity of each water molecule to form hydrogen bonds with four other water molecules permits an orderly arrangement of molecules. Although the individual hydrogen bonds are weak, collectively they give remarkable cohesiveness to the water molecules. In the liquid state, not all possible hydrogen bonds have formed, but in ice the ordering is at its maximum, with every molecule bonded to four others, each oriented as far as possible from the other three. The result is a rather open three-dimensional lattice, less dense than the liquid water on which it floats.

The high degree of cohesiveness between water molecules due to hydrogen bonding has other important consequences. Water's unusually high surface tension, melting point, boiling point, heat of vaporization, and heat capacity result from the characteristic hydrogen bonding pattern of water.

Because of the polarity of the water molecule, it is an excellent solvent for many chemicals. Ionic substances and nonionic polar substances dissolve easily in water, because of the electrostatic attraction between them and the charged parts of the water molecules. Electrically neutral and nonpolar substances are not soluble in water.

Some simple organic chemistry Molecules may be classified into two types: organic and inorganic. *Organic* compounds always contain the element carbon and are often large and complex; they are the principal materials that make up living systems. All other compounds are said to be *inorganic*; these molecules are usually smaller and less complex. Carbon dioxide, because of its relative simplicity, is usually considered inorganic, despite its carbon content. Many of the inorganic materials, notably water, carbon dioxide, and molecular oxygen, are also vital components of living things.

Carbon, with its covalent bonding capacity of four, commonly forms bonds with hydrogen, nitrogen, and other carbon atoms. The carbon atoms usually join in long chains or rings, forming molecules that may be very complex. Four major classes of organic molecules found in living organisms are carbohydrates, fats, proteins, and nucleic acids.

Although these four classes of molecules differ in structure and function, they are similar in that all are built up of many simpler "building-block" molecules bonded together. In each case these building-block molecules are combined by the removal of water molecules in *condensation reactions*. In such a reaction a hydrogen atom is removed from the end of one building-block molecule and a hydroxyl (OH) group from the end of a second molecule. The two building-block molecules are now joined together, and a molecule of water has been formed:

block molecules: glycerol and fatty acids. Each glycerol molecule can combine with three fatty acids by condensation reactions. The resulting fat may be *saturated* (having the maximum number of hydrogen atoms) or *unsaturated* (having at least one double bond between carbon atoms). The *phospholipids* are derived from the fats. In the most common ones, a group containing phosphate and nitrogen replaces the third fatty acid. Because this nitrogen and phosphate group is strongly polar while the rest of the molecule is nonpolar, one end of the phospholipid molecule is soluble in water while the other is not. This property of phospholipids makes them important constituents of the cell membranes. Another major group of lipids is the *steroids*, complex molecules made up of four interlocking rings of carbon atoms, whose physical properties are similar to those of the other lipids.

The basic building-block molecules of the *proteins* are the *amino acids*. An amino acid has

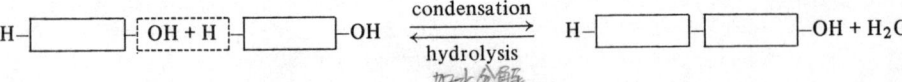

Condensation reactions are reversible; the complex organic molecules can be *hydrolyzed* into the simpler building-block molecules.

The basic building-block molecules of the *carbohydrates* are the simple sugars, or *monosaccharides*. Biologically important monosaccharides are the six-carbon sugars glucose, fructose, and galactose, which are *isomers* of each other; i.e. they have the same molecular formula, $C_6H_{12}O_6$, but differ in their molecular structure and biological properties. When two simple sugars are bonded together by a condensation reaction, a double sugar, or *disaccharide*, is formed. When many simple sugars are bonded together in long chains by the same kinds of condensation reactions, a *polysaccharide* is formed. Starch, glycogen, and cellulose are examples of polysaccharides formed by the bonding together of glucose molecules. Starch is the principal carbohydrate storage form for plants, glycogen for animals. The carbohydrates are an important energy source for all organisms.

The *lipids* consist of the fats and fatlike substances, which tend to be insoluble in water and soluble in many organic solvents. The neutral *fats* are important energy-storage compounds. Fats are made up of two different building-

an amino group (NH_2), a carboxyl group (COOH), and an R group attached to the same carbon atom. It is the different R groups, or side chains, that determine the uniqueness of each of the 20 amino acids. The R groups may be polar, nonpolar, or ionic and thus vary in their chemical properties. The amino acids are bonded together to form a protein by condensation reactions between the amino group of one amino acid and the carboxyl group of another. The resulting bond is the *peptide* bond and the chains produced are *polypeptide* chains. The *primary structure*, which determines the unique properties of each protein, is the sequence and type of amino acids making up the polypeptide chains. Because hydrogen bonds form between the nitrogen of one amino acid and the oxygen of another, the chain tends to assume a stable regular shape known as the *secondary structure* of the protein. These regular molecules may in turn be folded into complicated globular shapes by weak attractions between the different R groups present in the chain, thus forming the *tertiary structure* of the protein. It is therefore the sequence of the amino acids with their various R groups that is of critical importance in determining the tertiary structure. Some globular proteins are made up of two or more independently folded polypeptide chains held together by weak bonds; the way these chains fit together determines the *quaternary structure*. Because the conformation of a protein depends on weak bonds (which are very sensitive to temperature and pH), it is easily altered. Alteration in the three-dimensional structure of a protein can lead to a change in its biological function.

The *nucleic acids* are the organic material that makes up the genes and controls the structural characteristics of living things. The building-block unit of the nucleic acids is the *nucleotide*, which is made up of a five-carbon sugar attached by covalent bonding to a phosphate group and to a nitrogen-containing base:

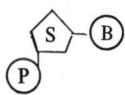

The nucleotide units are joined together through condensation reactions between the sugar of one nucleotide and the phosphate group of the next nucleotide in the sequence. Thus a long chain of alternating sugar and phosphate groups is established, with the nitrogenous bases oriented as side groups. There are four different nucleotides in each nucleic acid; it is the different sequences of the nucleotides that encode the hereditary information. The two types of nucleic acids, DNA and RNA, differ in their basic sugar molecule, nucleotide makeup, and the number of strands in the molecule.

Chemical reactions   Many different types of chemical reactions take place within living organisms. Chemical reactions that release free energy are *exergonic* reactions; reactions that require the addition of free energy are *endergonic*. In living systems an exergonic reaction is usually coupled with an endergonic reaction; some of the energy released by the exergonic reaction provides the energy to drive the endergonic reaction.

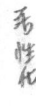

Although exergonic reactions are spontaneous, they may require some extra energy, or *activation energy*, to initiate the reaction. Once started, the reaction proceeds on its own and energy is released. This need for activation energy ordinarily prevents high-energy compounds such as lipids and carbohydrates from breaking down, and makes them stable.

Chemical reactions can be speeded up by heat, by increasing the concentrations of the reactants, or by providing an appropriate *catalyst*. In living systems the catalysts are *enzymes*— globular proteins that act to reduce the activation energy but are unchanged at the end of the reaction. The key to an enzyme's activity lies in the distinctive shape created by the precise folding of its polypeptide chain(s). Most enzymes are specific and can interact only with those reactants, or *substrates*, that fit spatially and are chemically compatible with the *active site* of the enzyme. During the reaction the enzyme and its substrate become temporarily bonded to form an unstable enzyme-substrate complex; this complex then separates, releasing the products and the unchanged enzyme. Enzymes may act in a number of ways; they may increase the rate of collision of the substrate molecules, orient the collisions to make them more effective, or enhance the reactivity of the substrates.

Since the formation of the enzyme-substrate complex requires the enzyme and its substrate to be complementary, it follows that anything that alters the shape of the enzyme will alter its activity. In addition to temperature and pH, many chemical substances can mask, block, or alter the shape of the enzyme and its active site. *Inhibitors* decrease enzyme activity either by competing with the substrate molecules for the active site and blocking it (competitive inhibitor), or by binding to a second site on the enzyme and physically blocking the active site or inducing a conformational change in the enzyme molecule (noncompetitive inhibitor).

QUESTIONS

Testing recall

Decide whether the following statements are true or false, and correct the false statements.

1 The isotope C^{14} is heavier than C^{12} because its atomic nucleus contains two additional protons.

2 The number of electrons at the outermost energy level determines the chemical properties of an atom.

12 • CHAPTER 2

3 Ions are formed when an atom loses or gains electrons.

4 When a solution is acidic, its pH is greater than 7.

5 The orderly three-dimensional arrangement of molecules in pure water is due primarily to the formation of ionic bonds between water molecules.

6 The marked polarity of water molecules makes water an excellent solvent for many important classes of chemicals.

7 Phospholipids are molecules with hydrophilic and hydrophobic portions.

8 Adding hydrogen to an unsaturated fat will make it more saturated.

9 Chemical reactions involving the combination of smaller building-block molecules with the removal of water molecules are hydrolysis reactions.

10 The bond between two adjacent amino acids in a protein molecule is a peptide bond.

11 Chemical reactions that require a net input of free energy to proceed are endergonic reactions.

12 All enzymes are proteins.

Structural diagrams for representatives of the four classes of organic molecules, or their component building-block molecules, are shown below. Match each molecule to the class to which it belongs.

a carbohydrate   c protein
b lipid   d nucleic acid

13
$$\text{glucose ring structure with CH}_2\text{OH, OH, H groups}$$

14
$$H_3N^+ - \underset{CH_3}{\underset{|}{C}} - \underset{O^-}{\overset{O}{\overset{\|}{C}}}$$

15
$$^+NH_3 - CH_2 - CH_2 - O - \underset{O}{\underset{|}{P}}(=O) - O - CH_2 - CH(O-C(=O)-CH_2CH_2CH_2CH_2CH_2CH_2CH_2CH_3) - CH_2 - O - C(=O) - CH_2CH_2CH_2CH_2CH_2CH_2CH_2CH_3$$

16
$$HO-\underset{CH_2CH_2CH_2CH_2CH_3}{\overset{O}{\overset{\|}{C}}}$$

17
$$\begin{array}{c} H \\ | \\ H-C-OH \\ | \\ C=O \\ | \\ HO-C-H \\ | \\ H-C-OH \\ | \\ H-C-OH \\ | \\ H-C-OH \\ | \\ H \end{array}$$

18
$$\begin{array}{c} H \\ | \\ H-C-OH \\ | \\ H-C-OH \\ | \\ H-C-OH \\ | \\ H \end{array}$$

19
$$\text{pyrimidine ring with HN, O=, CH}_3, NH$$

Testing knowledge and understanding

Choose the one best answer.

1. The chemical properties of an atom are primarily determined by the
   a number of protons.
   b number of electrons.
   c number of neutrons.
   d atomic weight.
   e number of isotopes.

2. A nonpolar covalent bond occurs
   a when one atom has a greater affinity for electrons than another.
   b when the constituent atoms attract the electrons equally.
   c when an electron from one atom is completely transferred to another atom.
   d between atoms whose outer energy levels are complete.
   e when the molecule becomes ionized.

3. Which one of the following kinds of biologically important bonds requires the most energy to break it?
   a van der Waals interaction
   b covalent bond
   c hydrophobic interaction
   d hydrogen bond
   e ionic bond

4. The cohesiveness among water molecules is due largely to
   a hydrogen bonds.
   b polar covalent bonds.
   c nonpolar covalent bonds.
   d hydrophobic interactions.
   e van der Waals interactions.

5. The carbon atom can form so many different chemical compounds because
   a its unstable nucleus easily gives up neutrons.
   b its outer energy level contains four electrons.
   c its electron shells are stable.
   d it can form both ionic and covalent bonds.
   e it tends to give up electrons to electron acceptors.

6. Compounds with the same atomic content but differing structures and properties are called
   a isotopes.
   b isomers.
   c ionic compounds.
   d polar covalent compounds.
   e nonpolar covalent compounds.

7. Two classes of organic compounds typically provide energy for living systems. Representatives of these two classes are
   a fats and amino acids.
   b amino acids and glycogen.
   c amino acids and ribose sugars.
   d fats and polysaccharides.
   e nucleic acids and phospholipids.

8. The chemical formula

$$\begin{array}{c} H \\ \diagdown \\ H \end{array} N - \begin{array}{c} H \\ | \\ C \\ | \\ R \end{array} - \begin{array}{c} O \\ \diagup\diagup \\ C \end{array} \longrightarrow \begin{array}{c} R \\ | \\ N - C \\ | \\ H \end{array} \begin{array}{c} OH \\ \diagup \\ - C \\ \diagdown\diagdown \\ O \end{array}$$

represents the product of a condensation reaction between
   a a fatty acid and an amino acid.
   b two fatty acids.
   c two amino acids.
   d an amino acid and an alcohol.
   e a fatty acid and an alcohol.

9. If two five-carbon sugars are combined to form a disaccharide molecule with ten carbons, how many *hydrogen* atoms will it have?
   a 10
   b 12
   c 18
   d 20
   e 22

10. Plants commonly store carbohydrates for an energy source as
    a glycogen.
    b starch.
    c cellulose.
    d sucrose.
    e fat.

For questions 11–13

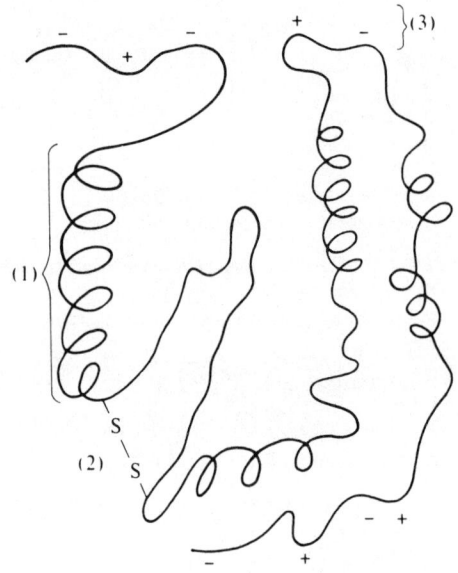

11 The drawing represents a protein molecule. The region labeled (1), taken alone, illustrates which level of protein structure?

   a primary
   b secondary
   c tertiary
   d quaternary

12 In the drawing the bond labeled (2) is

   a a covalent bond.
   b a hydrogen bond.
   c an ionic bond.
   d a hydrophobic bond.
   e a van der Waals interaction.

13 The region of the protein molecule labeled (3) is

   a hydrophobic.
   b hydrophilic.
   c nonpolar.
   d α-helical.

14 Both DNA and RNA

   a are single-stranded molecules.
   b contain the same four nucleotide bases.
   c are polymers of amino acids.
   d have the same five-carbon sugar.
   e contain phosphate groups.

15 The reaction A + B ⟶ C + D is exergonic. Which one of the following procedures would *not* be effective in accelerating the rate of the reaction?

   a increasing the concentration of C and D
   b heating A and B together
   c increasing the pressure on A and B
   d providing an appropriate catalyst

16 Which one of the following statements is true of enzymes?

   a Enzymes lose some or all of their normal activity if their three-dimensional structure is disrupted.
   b Enzymes are composed of ribose, phosphate, and a nitrogen-containing base.
   c The activity of enzymes is independent of temperature and pH.
   d Enzymes provide the activation energy necessary to initiate a reaction.
   e An enzyme acts only once and is then destroyed.

Questions 17–19 refer to the following situation:

The enzyme succinic dehydrogenase normally catalyzes a reaction involving succinic acid. Another substance, malonic acid, sufficiently resembles succinic acid to form temporary complexes with the enzyme, although malonic acid itself cannot be catalyzed by succinic dehydrogenase.

17 In this example succinic acid is

   a the substrate.   c an inhibitor.
   b the active site.   d the product.

18 Malonic acid is

   a the substrate.
   b a competitive inhibitor.
   c a positive modulator.
   d a negative modulator.
   e a coenzyme.

For question 19

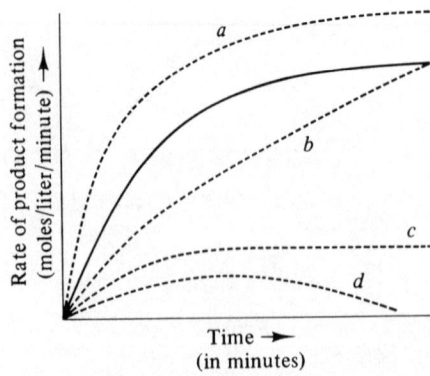

—— curve showing rate of product formation without the presence of malonic acid

---- possible curves showing rate of product formation with presence of malonic acid

SOME SIMPLE CHEMISTRY • 15

19 The solid line shows the reaction rate when a certain amount of the enzyme is added to 1 M succinic acid. Then the same amount of enzyme is added to 1 M succinic acid, but 0.5 M malonic acid is also present. Which of the dotted-line curves best shows the new rate of succinic acid catalysis?

20 Complete the following chemical reactions:

a

$$H_2N-\underset{\underset{H}{|}}{\overset{\overset{H}{|}}{C}}-C\overset{\overset{O}{\|}}{\underset{OH}{\diagdown}} \quad + \quad H_2N-\underset{\underset{CH_2}{|}}{\overset{\overset{H}{|}}{C}}-C\overset{\overset{O}{\|}}{\underset{OH}{\diagdown}} \xrightarrow{\text{condensation}}$$

(with phenol -OH group on CH₂)

b (triglyceride structure) + 3 H₂O $\xrightarrow{\text{hydrolysis}}$

c (two glucose rings) + $\xrightarrow{\text{condensation}}$

For further thought

1 Discuss the chemical nature and mechanism of action of enzymes. Include in your answer the terms substrate, active site, activation energy, products, specificity.

2 The enzyme pepsin is secreted by the stomach in a biologically inactive form called pepsinogen. Pepsinogen is changed into active pepsin in the stomach cavity. Activation results from the splitting off of 42 amino acids from one end of the pepsinogen molecule; the shorter polypeptide chain that remains is pepsin. The active pepsin catalyzes the hydrolysis of certain peptide bonds within the protein molecule. In terms of the conformation of the enzyme and its active site, explain how splitting off the 42 amino acids activates the enzyme.

Chapter 3

CELLS:
UNITS OF STRUCTURE AND FUNCTION

KEY CONCEPTS

1 The fundamental organizational unit of life is the cell.

2 All living things are composed of cells; all cells arise from pre-existing cells.

3 A living cell is an extraordinarily complex unit with an intricate internal structure; its activities are precisely integrated and controlled.

4 A cell's interaction with its environment is a crucial one; materials necessary for life must be obtained from the environment, and waste products must be released into it.

5 There are two fundamental types of cells, procaryotic and eucaryotic cells. Procaryotic cells lack a membrane-enclosed nucleus, as well as other intracellular membranous organelles present in eucaryotic cells.

OBJECTIVES

After studying this chapter and reflecting on it, you should be able to

1 Cite the contributions of each of the following scientists to our knowledge of cells and cell structure: Antoni van Leeuwenhoek, Robert Hooke, Matthias Schleiden, Theodor Schwann, Rudolf Virchow, Louis Pasteur, Camillo Golgi.

2 Describe the roles of the light microscope, transmission electron microscope, and scanning electron microscope in developing our current understanding of the details of cell structure.

3 Describe the Davson-Danielli and unit-membrane models of the plasma membrane, and explain why the fluid-mosaic model has superseded them.

4 Relate the permeability of the plasma membrane to its structure as envisioned in the fluid-mosaic model.

5 Explain why diffusion is driven by an increase in entropy of a system whereas exergonic chemical reactions are usually driven by a loss of heat energy.

6 Distinguish between the processes of active and passive transport. Indicate the role of these processes in the life of a cell.

7 Differentiate between osmosis and diffusion.

8 Define the terms hypotonic, hypertonic, and isotonic and explain why osmotic relationships are of critical importance in the life of the cell and organism.

9 Describe the formation of the plant cell wall and its significance in the life of the plant.

10 Describe the structure and give the function of the following organelles: nucleus, chromosomes, nucleoli, endoplasmic reticulum (rough and smooth), ribosomes, lysosomes, Golgi apparatus, vacuole, centriole, plastids, mitochondrion, peroxisome, cilia, flagella. Indicate which of these occur in plant cells and which in animal cells.

11 Discuss the similarities found in the structures of centrioles, basal bodies, cilia, and flagella.

12 List the differences between procaryotic and eucaryotic cells.

13 Classify and give the function of each of the major categories of plant and animal tissues.

SUMMARY

The fundamental organizational unit of life is the *cell*, the study of which dates from the work of Hooke in the mid-1600s. The mid-1800s saw the formulation of the *cell theory*, which says that all living things are composed of cells and that all cells arise from pre-existing cells, that there is no spontaneous creation of life. Proof of this theory of *biogenesis* came from Pasteur's work with microorganisms.

Most cells are very small, and have a large ratio of surface area to volume. Cells must exchange materials (e.g. oxygen, nutrients, wastes) with the surrounding environment, and these materials must move in and out across the surface of the cell. As cells grow larger, the volume increases much more rapidly than the surface area. Thus, a small cell has a more favorable ratio of surface area to volume for the exchange of materials to support itself than does a large cell.

The cell membrane Cells are bounded by a *plasma membrane* composed of lipids and proteins, with many small pores. In the late 1930s Danielli and Davson formulated a model of the cell membrane with two layers of phospholipids sandwiched between two continuous layers of protein. With support from early electron microscopy and X-ray diffraction studies, Robertson proposed this model, which he termed the *unit membrane*, as the basic structure for all cellular membranes.

Recent research, however, has provided evidence that cannot be well explained by the unit-membrane model. An alternative model was proposed in 1972 by Singer and Nicolson. Called the *fluid-mosaic model*, it envisions a bilayer of phospholipids with their hydrophilic heads oriented towards the surfaces of the membrane and their hydrophobic tails towards the interior. The proteins are distributed in a mosaic pattern, both on the surfaces (the *extrinsic proteins*) and in the interior of the membrane (the *intrinsic proteins*). The membrane structure is not static; some molecules can move, though others appear to be anchored in place. The pores are thought to be bounded by protein; the distinctive properties of the R groups of the amino acids in these proteins make the pores selective as to what can move through them.

The cell membrane is an active part of the cell, playing a complex, dynamic role in life processes. It regulates the movement of materials between the ordered interior of the cell and the outer environment. The general rule governing the movement of materials is that the net movement of particles of a particular substance is from regions of greater free energy (where there is an orderly, improbable arrangement) to regions of less free energy (where there is a disorderly, more probable arrangement) of that substance. This movement of particles is called *diffusion*. The plasma membrane is *differentially permeable*; it allows particles of some substances to pass through while restricting others. A membrane that is permeable to water but not to solute is said to be *semipermeable*. The movement of water through a semipermeable membrane is called *osmosis*.

If two different solutions are separated by a semipermeable membrane, under constant conditions of temperature and pressure the net movement of water will be from the solution with fewer dissolved particles per unit volume to the solution with more dissolved particles per unit volume; i.e. from the solution with the lower osmotic concentration to the solution with the higher osmotic concentration. Water will continue to move in response to the concentration difference until an equilibrium is reached; at equilibrium the hydrostatic pressure (and hence free energy) in the solution with the higher osmotic **concentration becomes so great** that water molecules begin to be forced back through the membrane as fast as other water molecules are moving in. When the system reaches this *dynamic equilibrium*, the hydrostatic pressure exerted by the solution with the higher osmotic concentration is known as *osmotic pressure*, the pressure that must be exerted on a solution to keep it in equilibrium with water when the two are separated by a semipermeable membrane. The osmotic pressure of a solution is a measure of the tendency of water to move into it by osmosis. Under constant conditions of temperature and pressure, water will move from the solution with the lower osmotic pressure to the solution with the higher osmotic pressure when the two are separated by a semipermeable membrane.

Because the plasma membrane is differentially permeable, the processes of osmosis and diffusion are fundamental to cell life. Cell membranes are relatively permeable to water and to certain simple sugars, amino acids, and lipid-soluble substances; they are relatively impermeable to polysaccharides, proteins, and other very large particles. Their permeability to small particles varies, but in general uncharged particles cross more rapidly than charged ones.

Although some molecules enter or leave the cell by simple diffusion, this does not explain completely how the plasma membrane regulates the passage of materials into and out of the cell. The bilipid layer presents a barrier to substances insoluble in lipids; some mechanism must be present in the membrane to transport these substances through it. Apparently, some of the protein components of the membrane function as transport agents or carrier molecules, called *permeases*. We do not know exactly how permeases transport their substrates through membranes, but two different types of carrier-mediated transport can be distinguished: facilitated diffusion and active transport. In *facilitated diffusion*, or passive transport, substances move with the concentration gradient and no energy is required. In *active transport* substances are moved against a concentration gradient; hence the cell must expend energy.

Sometimes substances are taken into the cell by an active process called *endocytosis*, in which a substance is enclosed in a membrane-bound vesicle pinched off from the cell membrane. The reverse sequence, in which materials within vesicles are conveyed to the surface of the cell and discharged, is called *exocytosis*.

The cell membrane cannot completely regulate the exchange of materials. A cell in a medium that is *hypertonic* (higher osmotic concentration) relative to it tends to lose water and shrink. Conversely, a cell in a *hypotonic* medium (lower osmotic concentration relative to it) tends to gain water and swell, and may even burst. A cell in an *isotonic* medium (osmotic concentrations in balance) neither gains nor loses appreciable water.

Cell walls and coats Since all protoplasm has a higher osmotic concentration than fresh water, all freshwater organisms live in a hypotonic environment and face the problem of taking in excessive water by osmosis. Unless the cell has some mechanism for expelling excess water, or special structures that prevent excess swelling, it may burst. The plant *cell wall*, an example of such a structure, is located outside the membrane and is composed mainly of thin fibrils of cellulose. The *primary wall*, the first portion of the wall laid down by a growing cell, consists of a loose network of fibrils. Cells that form the harder, woody portions of the plant add further layers, forming a thicker, more compact *secondary wall*. Adjacent cells are bound together by an intercellular layer known as the *middle lamella*; it is made up of a complex polysaccharide called *pectin*.

Fungi and bacteria also have cell walls, but unlike those of true plants they are not made of cellulose but rather of other complex polysaccharide molecules. The presence of the cell wall enables the cells of plants, fungi, and bacteria to exist in hypotonic media without bursting. In such media water will tend to move into the cell, causing it to swell and building up *turgor pressure* against the wall. The wall of a mature cell can usually be stretched very little; equilibrium is reached when it cannot hold any more water.

Most animal cells have a cell coat or *glycocalyx* composed of carbohydrates covalently bonded to protein or lipid molecules in the plasma membrane. The glycocalyx functions in cell recognition and in contact inhibition, and has recognition sites for interaction between the cell and important molecules.

The nucleus One of the largest and most conspicuous structural areas within *eucaryotic* cells is the membrane-bound nucleus. The cells of bacteria and blue-green algae, which lack a membrane-bound nucleus, are termed *procaryotic*. The following discussion concerns only eucaryotic cells, which may be divided into two major structural subdivisions: the nucleus and the *cytoplasm*.

The *nucleus* is the control center of the cell. The genetic material (*DNA*) contained in the *chromosomes* has the information for directing the cell's life processes, including cellular reproduction, cellular differentiation, and metabolic activities. The nucleus also contains one or more *nucleoli*, where ribosomal *RNA* is synthesized and combined with proteins before moving into the cytoplasm to become part of the ribosomes. Separating the nucleus from the cytoplasm is a double *nuclear membrane* interrupted at intervals by pores at points where the outer and inner membrane are continuous. Exchange of materials between the nucleus and the cytoplasm through the pores is highly selective. The nuclear membrane is continuous at places with the endoplasmic reticulum.

Cytoplasmic organelles The *endoplasmic reticulum* (ER) is a complex network of cytoplasmic membranes that form a system of membrane-enclosed spaces, some of which are interconnected. Sometimes the membranes of the ER are "rough," lined on their outer surfaces by small particles called *ribosomes*; when no ribosomes are present, the ER is "smooth." The ribosomes are the site of protein synthesis. The ER functions both as a passageway for intracellular transport and as a manufacturing surface. Some of the protein molecules making up the ER membranes act as enzymes for the cell's biochemical pathways.

The endoplasmic reticulum also has connections with the *Golgi apparatus*, which consists of stacks of membrane-bound vesicles that function in the storage, modification, and packaging of secretory products.

Membrane is thought to flow through the cellular membrane system as follows: nuclear envelope ⟶ rough ER ⟶ smooth ER ⟶ Golgi apparatus ⟶ secretory vesicles ⟶ plasma membrane.

Located within the cytoplasm are many other organelles. The *mitochondria* are the powerhouses of the cell; chemical reactions within the mitochondria provide energy for the activities of the cell. Each mitochondrion is a double-membrane vesicle; the outer membrane is smooth while the inner one has folds that project into the interior. *Lysosomes* are membranous sacs that function as storage vesicles for powerful digestive enzymes; they may act as the cell's digestive system, hydrolyzing materials taken in by endocytosis. *Peroxisomes* are also membranous sacs of enzymes; these are oxidative rather than digestive. Most plant cells have large membranous organelles called *plastids*. The two principal categories of plastids are the colored *chromoplasts* and the colorless *leucoplasts*. Chloroplasts are chromoplasts that contain the green pigment chlorophyll; they capture the energy of sunlight and utilize it in the manufacture of organic compounds. The leucoplasts' primary function is storage of starch, oils, or protein granules. Membrane-enclosed, fluid-filled spaces termed *vacuoles* are found in many cell types, and perform a variety of functions. Most mature plant cells have a large central vacuole occupying much of the volume of the cell; it plays an important role in maintaining the turgidity of the cell and in the storage of important substances, including wastes.

Microtubules and *microfilaments* appear to function in intracellular movement and to support the cell. Microtubules also form the spindle of dividing cells and are essential components of centrioles, cilia, and flagella. The *centrioles* are hollow cylindrical bodies located just outside the nucleus of most animal cells; they help organize the microtubular spindle for cell division. *Cilia* and *flagella* are hairlike projections from the cell's surface that move the cell or move materials across the cell's surface. Inside the stalk of cilia and flagella is a 9 + 2 arrangement of microtubules. At the base of the stalk is the *basal body*, whose structure is the same as that of the centriole.

Procaryotic cells lack the internal membranous organelles discussed above, but do have ribosomes and a *nucleoid* containing a circular chromosome of DNA. Some cells have flagella but these lack microtubules.

Plant and animal tissues The bodies of multicellular organisms are organized on the basis of *tissues*, organs, and systems. All plant tissues can be divided into two major categories: meristematic tissue (growth tissue—undifferentiated cells capable of dividing) and permanent tissue (mature differentiated cells). Regions of *meristematic tissue* are found at the growing tips of roots and stems (*apical meristems*) and, in many plants, in areas towards the periphery of the roots and stems (*lateral meristems*). The permanent tissues fall into three subcategories: surface tissues, fundamental tissues, and vascular tissues. The surface tissues (*epidermis, periderm*) form the protective outer covering of the plant body. Most of the fundamental tissues are simple tissues; i.e. each is composed of a single cell type. There are four types of fundamental tissues: *parenchyma* (thin-walled, loosely packed cells found throughout the plant body), *collenchyma* (supportive tissue whose cell walls are characteristically thickened at the "corners"), *sclerenchyma* (supportive tissue with very thick, heavily lignified cell walls), and *endodermis* (a single layer of cells surrounding the vascular cylinder of roots). The vascular or conductive tissue is characteristic of higher plants and consists of two principal types of complex tissue: xylem and phloem. *Xylem* supports the plant and transports water and dissolved minerals upward. *Phloem* conducts organic materials up and down the plant body. The body of higher land plants is divided into two major parts: the root and the shoot.

Animal tissues are divided into four categories: epithelium, connective tissue, muscle, and nerve. *Epithelial tissue* covers or lines the internal and external surfaces of all free body surfaces. Epithelial cells may also become specialized as gland cells, hair, and other surface derivatives. *Connective tissue*, composed of cells imbedded in an extensive extracellular matrix, connects, supports, or surrounds other tissues or organs; examples of connective tissue are blood and lymph, connective tissue proper, cartilage, and bone. The three types of muscle tissue (smooth, skeletal or striated, and cardiac) consist of cells specialized for contraction and are responsible for most movement in higher animals. *Nervous tissue* is highly specialized for the ability to respond to stimuli. The functional combination of nerve and muscle tissue gives animals their characteristic ability to move rapidly in response to stimuli.

QUESTIONS

Testing recall

This chapter introduces many new names and terms that are part of the basic biological vocabulary. It is essential for you to learn these words now, for you will encounter them again and again.

Listed below are eleven scientists. Match their names with their contributions.

a Danielli and Davson
b Hooke
c Pasteur
d Robertson
e Schleiden and Schwann
f Singer and Nicolson
g van Leeuwenhoek
h Virchow

1 formulated the cell theory

2 advanced the theory of biogenesis

3 proposed the lipoprotein-sandwich model of cell membrane

4 formulated the unit-membrane concept

5 developed the first microscope

6 disproved the theory of spontaneous generation

7 first used the term *cell*

8 formulated the fluid-mosaic model

It is important not to confuse the following pairs of terms. In one sentence, distinguish between the words in each pair.

9 osmosis – diffusion

10 hypertonic – hypotonic

11 osmotic concentration – osmotic pressure

12 unit-membrane model – fluid-mosaic model

13 facilitated diffusion – active transport
14 endocytosis – exocytosis
15 pinocytosis – phagocytosis
16 cell wall – cell coat

17 primary cell wall – secondary cell wall
18 procaryotic – eucaryotic
19 epithelium – epidermis

---

20 Work the crossword puzzle.

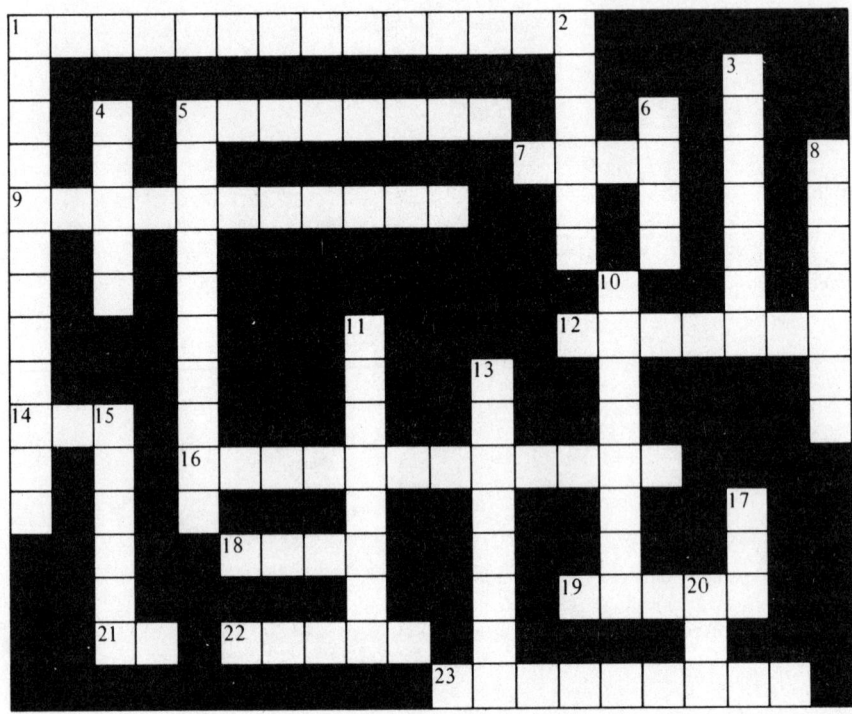

Across

1 long threadlike structures found in muscle fibers
5 protein-carrier molecules in the membrane
7 opening in the nuclear envelope
9 chromoplast containing chlorophyll
12 control center of cell
14 found in nucleus and ribosomes
16 hollow cylindrical structures found in spindle fibers and centrioles
18 prefix meaning "less than"
19 have a 9 + 2 arrangement of microtubules
21 abbreviation for the organelle involved in intracellular transport
22 prefix meaning "greater than"
23 organizes the microtubular spindle in animal cells

Down

1 powerhouses of cell
2 type of endoplasmic reticulum
3 occupies most of the volume of most mature plant cells
4 membrane-bound vesicles involved in packaging secretory products
5 contains oxidative enzymes
6 fundamental organizational unit of life
8 the fluid part of the cytoplasm
10 manufacture part of the ribosomes
11 cell's digestive system
13 site of protein synthesis
15 carrier-mediated transport that requires energy
17 contains the hereditary information
20 prefix meaning "equal to"

Listed below are ten substances or organelles found in many cells. Next to each, write E if the item is found only in eucaryotic cells, P if it is found only in procaryotic cells, and B if it is found in both.

21 nuclear membrane

22 ribosomes

23 nucleolus

24 flagella without 9 + 2 structure

25 DNA in chromosome

26 mitochondria

27 lysosome

28 plasma membrane

29 chlorophyll

30 endoplasmic reticulum

Match the following plant and animal tissues with their characteristics.

a collenchyma   g muscle
b connective   h nervous
c endodermis   i parenchyma
d epidermis   j phloem
e epithelium   k sclerenchyma
f meristematic   l xylem

31 thin-walled, loosely packed, unspecialized plant tissue

32 protective outer covering of the plant body

33 blood and bones are examples

34 undifferentiated plant cells capable of dividing

35 conducts organic materials up and down the plant body

36 specialized for contraction

37 uniformly thick-walled supportive tissue

38 conducts water and dissolved materials upward in the plant body

39 specialized for stimulus reception and conduction

40 lines the inner and outer surfaces of the animal body

Testing knowledge and understanding

The solutions in the two arms of the U tube are separated at the bottom of the tube by a differentially permeable membrane. At the beginning of the experiment the volumes in both arms are the same, and the level of the liquid is therefore at the same height. The membrane is permeable to water and to sodium chloride, but *not* to glucose. The apparatus is allowed to stand for three days.

Initial set-up

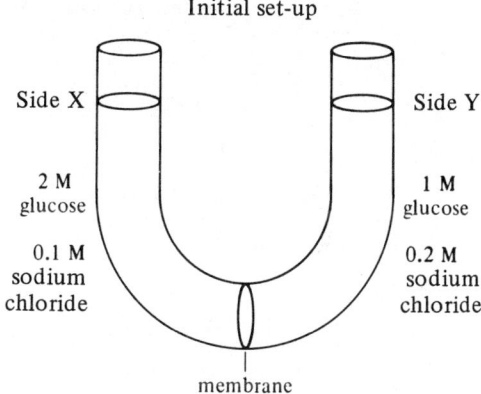

Questions 1-10 refer to the above diagram. For each one, select the most appropriate phrase using the following key:

a Both the *statement* and *reason* are correct.
b The *statement* is correct, but the *reason* is incorrect.
c The *statement* is incorrect, but the *reason* is a fact or principle.
d Both the *statement* and *reason* are incorrect.

1 The sodium chloride solution on side X will become more concentrated and that on side Y less concentrated *because* a substance tends to diffuse from regions of lower concentration to regions of higher concentrations of that substance.

2 The concentrations of the glucose solutions on sides X and Y will remain unchanged *because* the membrane is impermeable to glucose and it cannot diffuse from one side to the other.

3 The concentration of sodium chloride on side X will eventually equal that on side Y *because* sodium and chloride ions will move by diffusion from one side to the other until a uniform density is reached.

4 The concentration of glucose on side X will decrease and that on side Y increase *because* water molecules will diffuse through the membrane from side Y to side X by osmosis, thus lowering the glucose concentration on side X.

5 The fluid level will increase in side Y and decrease in side X *because* water molecules will move through the membrane from regions of higher to regions of lower concentration of water molecules.

6 The fluid level on side X will rise *because* the water molecules on that side at the beginning of the experiment have more free energy than those on side Y.

7 The net movement of water molecules will be from side X to side Y because water molecules will move from the solution with the lower osmotic pressure to the solution with the higher osmotic pressure when the two are separated by a semipermeable membrane.

8 Water molecules will move only from side Y to side X and not from side X to side Y because water molecules move only from regions of higher to regions of lower concentration.

9 The fluid level on side X will rise because the solution in side X had a higher osmotic pressure than the solution in side Y.

10 Water molecules will tend to move from side Y to side X because the net movement of water molecules will be from the solution with the lower to the solution with the higher osmotic concentration.

Choose the one best answer.

11 The figure shows two beakers interconnected by a tube partitioned by a membrane permeable to water but not to glucose. Beaker A contains a 2 percent glucose solution and beaker B contains a 4 percent glucose solution. Assuming uniform temperature and pressure, which one of the following statements best describes what will happen in this system?

a Water will move from A to B.
b Water will move from B to A.
c Water will move equally in both directions, so that there will be no net change in the system.
d Water will move in both directions but the net flow will be from A to B.
e Water will move in both directions but the net flow will be from B to A.

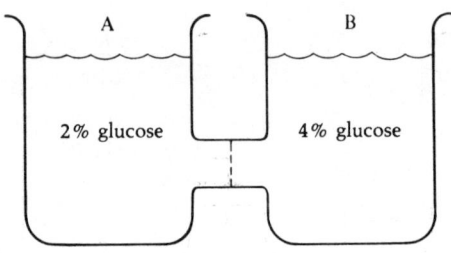

12 Suppose that instead of glucose, we dissolve 10^6 molecules of NaCl in beaker A and 10^6 molecules of sucrose in beaker B. Suppose further that the membrane in the connecting tube is impermeable to both NaCl and sucrose. What will happen in the system? (Select one of the answers listed under question 11.)

Questions 13–16 refer to the diagram of the plasma membrane.

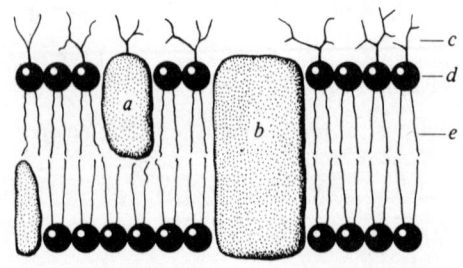

13 Of the five labeled parts of the model (*a, b, c, d,* or *e*), which one is now thought to play a critical role in cell recognition?

14 Which one of the labeled parts illustrates an integral globular protein that probably acts as a permease?

15 Which one of the labeled parts illustrates the electrically charged portion of a phospholipid molecule?

16 When hydrolyzed, substance B would yield

   a amino acids.
   b fatty acids.
   c monosaccharides.
   d nucleotides.
   e phospholipids.

17 A particular cell steadily secretes a protein into the surrounding medium. The secretion is released from the cell by exocytosis from membranous vesicles derived from the Golgi apparatus. The cell also functions as a storage depot for glycogen. The cell can be identified as

   a a bacterium.
   b a blue-green algal cell.
   c a cell from a terrestrial plant.
   d a cell from an animal.

18 In the following list, plant parts or products are paired with animal parts or products that are in some sense analogous to them. Which pair contains the *least functional* analogy?

   a epidermal cells – epithelial cells
   b bone – sclerenchyma
   c middle lamella – glycocalyx
   d starch – glycogen
   e xylem and phloem – blood vessels

19 The concentration of potassium in a red blood cell is much higher than it is in the surrounding blood plasma, yet potassium continues to move into the cell. The process by which potassium moves into the cell is called

   a osmosis.
   b simple diffusion.
   c facilitated diffusion.
   d active transport
   e pinocytosis.

20 The antibiotic streptomycin is thought to associate irreversibly with the ribosomes in bacteria and thus disrupt their normal functioning. In other words, this antibiotic destroys bacteria by

   a preventing the synthesis of proteins.
   b interfering with normal cell reproduction.
   c slowing energy production within the cell.
   d preventing transport within the endoplasmic reticulum.
   e interfering with materials entering and leaving the cell.

21 An organelle surrounded by a double membrane is the

   a lysosome.
   b Golgi apparatus.
   c chloroplast.
   d peroxisome.
   e ribosome.

22 Which one of the following pairs is mismatched?

   a nucleolus – site of ribosomal RNA synthesis
   b nucleus – DNA
   c lysosome – enzyme synthesis
   d cell membrane – lipid bilayer
   e smooth endoplasmic reticulum – lipid synthesis

For further thought

1 Distinguish between osmosis and diffusion and give a physical explanation for both. What is osmotic pressure? Discuss the implications of these processes for both an animal cell and a plant cell living in a hypotonic medium, a hypertonic medium, and an isotonic medium.

2 Describe a current model of the cell membrane and indicate the type of evidence upon which this model is based.

3 Describe the principal structural subdivisions of cells as seen with the electron microscope and indicate the functions of each. Compare plant, animal, and procaryotic cells with regard to these, where appropriate.

4 A 5 percent glucose solution is frequently administered intravenously to persons who have undergone surgery. Would you expect the 5 percent glucose solution to be hypertonic, hypotonic, or approximately isotonic to the blood? Give reasons for your answer.

5 Discuss the structure and function of microtubules and relate this to the various similarities and differences that exist among centrioles, basal bodies, cilia, and flagella.

Chapter 4

ENERGY TRANSFORMATIONS

KEY CONCEPTS

1. All systems, including living systems, have a natural tendency toward increasing disorder; the cell must constantly expend energy to maintain its highly organized state.

2. All cells must have a continuous supply of energy, since every energy transformation results in a reduction of the free (usable) energy of the system.

3. The ultimate energy source for most organisms is sunlight; green plants transform light energy into chemical energy, which can be used directly or passed to other organisms.

4. In living cells, almost all chemical syntheses and breakdowns are accomplished step by step, in a series of reactions each catalyzed by its own enzyme. The enzymes are often arranged in multienzyme systems, in which the product of one reaction becomes the substrate for the next.

5. The potential energy stored in complex organic molecules must be transformed into the energy of ATP—the universal energy currency of living organisms—in order to be used by the cell; this transformation must occur in every living cell.

6. The amount of energy that can be extracted from glucose if oxygen is present is considerably greater than if it is absent; consequently a plentiful supply of oxygen is essential for most organisms if their energy demands are to be met.

7. The metabolic breakdown of high-energy compounds is an inefficient process; more than half the available energy is lost as heat. Some animals have evolved mechanisms for retaining this heat and can thus maintain a uniformly high body temperature and metabolic rate.

OBJECTIVES

After studying this chapter and reflecting on it, you should be able to

1. Explain how the First and Second **Laws of Thermodynamics** relate to living organisms.

2. Explain how light energy is transformed into chemical energy in the light reactions of photosynthesis and what role is played by the accessory pigments in this process.

3. Define the terms *oxidation* and *reduction*, and explain how most oxidation-reduction reactions are carried out in living systems.

4. Compare the processes of cyclic and noncyclic photophosphorylation, and identify the products of each.

5. Give the structure of the ATP molecule and explain how it is formed and what role it plays in the transfer of energy.

6. Explain how the products of the light reactions are used to reduce CO_2 to form PGAL, and describe the fate of this PGAL.

7. Describe the process of photorespiration and indicate why it appears to be a wasteful process. Discuss the relationship between photosynthesis and photorespiration.

8. Sketch a chloroplast, including in your diagram the thylakoids, grana, and stroma. Indicate where the various photosynthetic reactions take place.

9. Describe the structural differences between C_3 and C_4 plants; explain the functional significance of the Kranz anatomy.

10. Describe and explain the significance of the processes of glycolysis and fermentation.

11. Describe the process of cellular respiration, including the main stages in the process, the role of oxygen, and the amount of energy produced.

12. Contrast the roles played by the electron acceptors NADP and NAD in metabolic processes.

13. Account for the critical importance of the respiratory electron-transport chain, and explain why cyanide and certain other poisons that block the chain are lethal.

14. Relate the structure of a mitochondrion to its function.

15. Cite differences between poikilothermic and homeothermic animals; explain how these differences relate to their activity level.

16. State the relationship between body size and metabolic rate, and explain its implications for small animals.

SUMMARY

Acquisition of energy in a usable form is a necessity for every living cell. According to the First Law of Thermodynamics, energy, which is the capacity to do work, can be converted from one form into another without any net gain or loss. Living cells draw primarily on chemical energy; they do work by using the energy stored in complex organic molecules. An organism must have an outside source of usable energy to replenish its supply of potential chemical energy, since, according to the Second Law of Thermodynamics, every transformation of energy results in a reduction of the *free* (usable) *energy* of the system.

Photosynthesis: The light reactions The ultimate energy source for most living things is sunlight, transformed by green plants into chemical energy in the process called *photosynthesis*. The green plants utilize the energy of light to combine carbon dioxide with water to form organic material (sugar) and oxygen. The summary equation for the photosynthetic process is

$$6CO_2 + 12H_2O + \text{light} \xrightarrow{\text{chlorophyll}} 6O_2 + C_6H_{12}O_6 + 6H_2O$$

Photosynthesis is a reduction reaction. *Reduction* is the addition of one or more electrons to an atom or molecule, while *oxidation* is the removal of electrons. Reduction stores energy; oxidation releases it. In biological systems many oxidation-reduction reactions involve the addition or removal of an electron derived from hydrogen. During photosynthesis two key processes occur. Light energy is trapped and stored, and hydrogen atoms are transferred from water to carbon dioxide to form carbohydrate.

Different wavelengths of light, especially red and blue light, are trapped by various pigment molecules organized in *photosynthetic units* within the chloroplasts. Two types of photosynthetic units occur in most plants; they constitute

Photosystem I and Photosystem II, respectively.

When a photon of light strikes a pigment molecule in Photosystem I and is absorbed, the energy is transferred to an electron, which is thereby raised to a high energy state. A specialized form of chlorophyll *a* eventually passes the energized electron to an acceptor molecule, X, which has a high affinity for electrons. X passes the electron to a series of acceptor molecules, each at a slightly lower energy level; each transfer is accomplished by an enzyme-catalyzed reaction. After being passed from molecule to molecule, the electron may return to the chlorophyll from which it started. Some of the energy released as the electron is eased down the energy gradient is used to synthesize the compound *ATP* from ADP and inorganic phosphate.

ATP is a universal energy currency used by cells to do all manner of work. It is composed of adenosine bonded to three phosphate groups in sequence: adenoisine $-$ ⓟ $\sim$ ⓟ $\sim$ ⓟ. If the terminal phosphate group is removed by hydrolysis, a large amount of energy is released and the compound *ADP* is left. New ATP can be synthesized from ADP and inorganic phosphate in an energy-demanding process called *phosphorylation*. When the energy necessary for ATP synthesis comes from light-energized electrons as they are returned to the chlorophyll molecules from which they originated, the process is termed *cyclic photophosphorylation*.

Another, more complicated process is carried on by most green plants. Again, light energy excites electrons in the pigments of Photosystem I, and the energized electrons are passed to electron acceptor X. X now donates the electrons to a different series of electron-transport molecules, the final acceptor being $NADP_{ox}$, which retains the electrons and is thus reduced ($NADP_{re}$). Photosystem I is now short of electrons; it has an electron "hole." Another light reaction occurs in Photosystem II, and the energized electrons are passed to an electron acceptor Q, which in turn passes them, via a series of transport molecules, step by step down an energy gradient; some of the energy released in the process is used in the synthesis of ATP. Eventually the electrons reach Photosystem I and fill the "hole." But Photosystem II is now short of electrons; the deficit is made good by electrons pulled from water. The splitting of water also produces free protons and molecular oxygen:

$$2H_2O \longrightarrow 4e^- + 4H^+ + O_2$$

Since the electrons are not passed in a circular chain in this process (some leave the system via $NADP_{re}$ and others enter from water), this series of reactions is termed *noncyclic photophosphorylation*; ATP, $NADP_{re}$, and O_2 are the end products.

Photosynthesis: The dark reactions in C_3 plants
The ATP and $NADP_{re}$ produced in the light reactions are used to reduce carbon dioxide to carbohydrate in a series of reactions called the *Calvin cycle*. First, a five-carbon sugar, ribulose diphosphate (RuDP), is combined with CO_2, a process known as carboxylation. The products are then phosphorylated by ATP and reduced by $NADP_{re}$ to form PGAL, a three-carbon sugar. Five-sixths of the PGAL produced is used to synthesize more RuDP; the remaining one-sixth may be used for synthesis of glucose.

Under certain conditions, however, RuDP is oxidized by means of the very same enzyme that under more favorable conditions would facilitate its carboxylation. This breakdown of RuDP is called *photorespiration*—a seemingly wasteful process, because no ATP is formed. Photorespiration predominates over photosynthesis when CO_2 levels are low and O_2 levels are high, but even at normal levels most photosynthetic production is undone by concurrent photorespiration.

Structure of leaves and chloroplasts The leaves of higher green plants are the principal organs of photosynthesis. The outer surfaces of a leaf are made up of epidermal tissue, covered with a waxy *cuticle* impermeable to water. The region between the upper and lower epidermis is filled with parenchyma cells making up the *mesophyll* region. The mesophyll cells have many chloroplasts and, being loosely packed, leave air spaces that communicate with the outside by tiny holes in the epidermis called *stomata*. Veins, which contain the xylem and phloem cells for transport, branch profusely within the mesophyll. The veins are usually surrounded by tightly packed parenchyma cells making up a *bundle sheath*.

Photosynthesis takes place within the chloroplasts. The pigments are precisely arranged within the membranes of flattened sacs called *thylakoids*. Disc-shaped thylakoids often lie close together in stacks called *grana*. The light reactions of photosynthesis (cyclic and noncyclic photophosphorylation) take place within the thylakoid membranes, while the dark reactions (Calvin cycle) take place in the colorless matrix (*stroma*) surrounding the thylakoids.

Photosynthesis: C₄ plants Some angiosperm plants of tropical origin have a distinct leaf structure known as *Kranz* anatomy. In Kranz plants the bundle-sheath cells have numerous chloroplasts (those of other plants usually do not), and mesophyll cells are clustered in a ring-like arrangement around the bundle sheath. Such plants can carry out photosynthesis under conditions of high temperature and intense light, when loss of water induces closure of the stomata; with the stomata closed, concentrations of CO_2 in the air spaces inside the leaf fall and those of O_2 rise. Under these conditions most plants (C_3 plants) have a net loss of CO_2 because of photorespiration, but Kranz plants (C_4 plants) do not, thanks to a special way of fixing CO_2 initially. They combine CO_2 with a three-carbon compound in the mesophyll cells to form a four-carbon compound that passes into the bundle-sheath cells, where the CO_2 is regenerated; they can thus maintain a CO_2 level in the bundle-sheath cells that allows carboxylation of RuDP in the Calvin cycle to predominate over its oxidation in photorespiration.

Cellular metabolism Cellular *metabolism* is a general term embracing all the myriad enzyme-mediated reactions of a living cell; it can be divided into two phases: *anabolism*, the building-up phase, and *catabolism*, the breaking-down phase. Photosynthesis is an anabolic process; light energy is stored in complex organic compounds such as glucose. Before this potential energy can be used by the cell to do work, it must be broken down in a series of chemical reactions and the energy transferred to ATP. The complete degradation of an energy-rich compound such as glucose to carbon dioxide and water involves many enzymatically controlled reactions.

Anaerobic metabolism The first series of reactions in the degradation of glucose is termed *glycolysis*; it is the breakdown of glucose to two molecules of pyruvic acid, with the production of two molecules of NAD_{re} and a net gain of two ATP molecules. This process, which is common to all living cells, is *anaerobic*; i.e. it does not require molecular oxygen.

The fate of the pyruvic acid depends on the oxygen supply. In the absence of sufficient O_2, the pyruvic acid may be reduced by NAD_{re} to form CO_2 and ethyl alcohol in some kinds of organisms, or lactic acid in others. The NAD_{ox} molecules formed in this reaction are then available to be reused in glycolysis. The process whereby the glycolytic pathway leads to the production of alcohol or lactic acid from pyruvic acid is called *fermentation*; it enables the cell to continue synthesizing ATP by the breakdown of nutrients under anaerobic conditions.

Aerobic metabolism Under *aerobic* conditions—when molecular oxygen is available—the pyruvic acid can be further oxidized, with the accompanying synthesis of ATP; this process is called *cellular respiration*. The process begins with the breakdown of the two pyruvic acid molecules to form two molecules each of acetyl-CoA, CO_2, and NAD_{re}. The two-carbon acetyl-CoA is fed into a complex circular series of reactions, the *Krebs citric acid cycle*. Here the acetyl-CoA combines with a four-carbon compound to form the six-carbon molecule citric acid. In subsequent reactions two carbons are lost as CO_2, leaving a four-carbon molecule that can combine with another acetyl-CoA and start the cycle over again. In the course of the cycle, a molecule of ATP is synthesized, and eight hydrogens are removed and picked up by carrier compounds (NAD_{ox} and FAD_{ox}). Since one molecule of glucose gives rise to two molecules of acetyl-CoA, two turns of the cycle occur for each molecule of glucose oxidized.

The final stage of respiration involves the passage of the hydrogen electrons from the carrier molecules down a "respiratory chain" of electron-transport molecules to oxygen, with which the hydrogen combines to form water. As the electrons are lowered step by step down the energy gradient, energy is released, and some of it is used to make ATP, a process called *oxidative phosphorylation*.

The total number of new ATP molecules produced by the complete metabolic breakdown of glucose is usually 36: two from glycolysis, two from direct synthesis in the Krebs cycle, and 32 from the electron-transport chain.

Anaerobic metabolism takes place in the cytoplasm, while aerobic respiration is confined to the mitochondria; the reactions of the Krebs cycle take place within the matrix of the inner compartment of the mitochondria, whereas the reactions of the electron-transport chain take place in the inner membrane. Evidence from studies of fragmented membranes indicates that the various acceptor molecules are structural components of the inner membrane and are arranged in clusters, each of which is a self-contained functional unit called a *respiratory assembly*.

Body temperature and metabolic rate Cellular respiration captures about 38 percent of the energy of glucose and converts it into ATP; the rest of the energy is released, mostly as heat. The vast majority of animals—termed cold-blooded or *poikilothermic* (of variable temperature)—and all plants promptly lose most of this heat to their environment. The body temperature of poikilotherms is determined by the environmental temperature. As their body temperature increases, their metabolic rate also increases (within limits); their activity is thus correlated with the environmental temperature.

A few animals, notably birds and mammals, maintain a constant high body temperature. Such animals are said to be warm-blooded or *homeothermic* (of uniform temperature). Their metabolic rate can accordingly be maintained at a uniformly high level, and they can remain active in cold weather. In both homeothermic and poikilothermic animals, as well as in plants, the normal metabolic rate is inversely related to body size; the smaller the organism, the higher the metabolic rate.

QUESTIONS

Testing recall

Underline the correct word or words to complete each statement.

1 "Energy can neither be created nor destroyed." This is a statement of the (First, Second) Law of Thermodynamics.

2 The energy contained in complex organic molecules is an example of (kinetic, potential) energy.

3 Reactant A in the reaction A + H ⟶ AH is (oxidized, reduced).

4 The reaction ADP + P$_i$ ⟶ ATP (stores, releases) energy.

5 The wavelengths of light most effective in driving photosynthesis are those in the (red, green) part of the visible spectrum.

6 The products of the light reactions necessary to drive the dark reactions are (ATP, NADP$_{re}$, O$_2$, CO$_2$).

7 Photorespiration (produces, destroys) RuDP.

8 The end product of glycolysis is (acetyl-CoA, pyruvic acid, lactic acid).

9 Alcoholic fermentation produces (CO$_2$, ethyl alcohol).

10 The reactions of the Krebs cycle take place within the (chloroplasts, mitochondria, cytosol).

11 Most of the ATP yield from the complete oxidation of glucose results from (glycolysis, the Krebs cycle, the electron-transport chain).

12 Animals whose body temperature varies with the environmental temperature are said to be (homeothermic, poikilothermic).

13 A cat has a lower metabolic rate than a (mouse, cow).

Of the basic processes of photosynthesis

 a cyclic photophosphorylation
 b noncyclic photophosphorylation
 c both light reactions
 d Calvin cycle
 e both light and dark reactions

which process (or group of processes) involves

14 fixation of CO$_2$

15 oxidation-reduction reactions

16 light-energized electrons

17 production of NADP$_{re}$

18 synthesis of PGAL

19 occurrence in thylakoid discs

20 occurrence in stroma of chloroplasts

21 splitting of water

22 production of ATP

23 chlorophyll as both the initial electron donor and the ultimate electron acceptor

Choose the one best answer.

24 The glycolytic pathway from glucose to pyruvic acid involves a lengthy series of different chemical reactions. Each of these reactions requires

a a molecule of ATP.
b a molecule of NAD.
c a molecule of ADP.
d a molecule of a specific enzyme.
e a molecule of NADP.

25 The internal membranes of an organelle taken from a eucaryotic cell are found to contain enzymes and coenzymes that carry out oxidative phosphorylation of ADP via a series of step-by-step reactions. The organelle is a

a chloroplast.   d Golgi apparatus.
b lysosome.   e mitochondrion.
c peroxisome.

26 The final electron acceptor in oxidative phosphorylation is

a O_2.   d NAD_{re}.
b H_2O.   e FAD_{re}.
c CO_2.

27 When a muscle cell is metabolizing glucose in the complete absence of molecular oxygen, which one of the following substances is *not* produced?

a PGAL   d lactic acid
b ATP   e acetyl-CoA
c pyruvic acid

28 To a living animal, which of the following compounds has the greatest amount of available energy per molecule?

a ATP   d CO_2
b ADP   e pyruvic acid
c H_2O

Testing knowledge and understanding

1 In the reaction NADH + H⁺ + FAD ⟶ NAD⁺ + FADH₂, which reactant will be oxidized as the reaction proceeds to the right?

a NADH   d NAD⁺
b FAD   e FADH₂
c H⁺

2 In biology a "limiting factor" is a condition or substance that, by its absence or short supply, limits the rate at which a biological process can proceed. Which one of the following would be *least* likely to be a limiting factor for photosynthesis?

a oxygen   d light
b carbon dioxide   e chlorophyll
c water

3 In photosynthesis the reduction of CO_2 to carbohydrate requires that both energy currency and a strong reducing substance (hydrogen donor) be in ample supply. The process of noncyclic photophosphorylation provides these requisites, using light energy to synthesize

a ADP and ATP.   d ADP and $NADP_{re}$.
b ATP and P700.   e P700 and P680.
c ATP and $NADP_{re}$.

4 If photosynthesizing green algae are provided with CO_2 synthesized with heavy oxygen (O^{18}), later analysis will show that all but one of the following compounds produced by the algae contain the O^{18} label. That one exception is

a PGA.   d RuDP.
b PGAL.   e O_2.
c glucose.

5 The main function of cyclic photophosphorylation in photosynthesis is to

a produce glucose.   d regenerate RuDP.
b decompose water.   e produce ATP.
c produce PGAL.

6 Which one of the following statements is *not* true of noncyclic photophosphorylation?

a Two different light-driven events are necessary if electrons are to be moved all the way from H_2O to NADP.

b The pigment molecules that trap light energy are built into the thylakoid membranes.

c There are at least two different places in the overall noncyclic pathway where energized electrons are passed energetically downhill via a series of intermediate electron-carrier substances.

d One of the products of noncyclic photophosphorylation that help make possible the dark reactions of the Calvin cycle is $NADP_{re}$.

e Some of the energy released during electron transport is used to drive the hydrolysis of ATP to ADP and inorganic phosphate.

7 Two different principal mechanisms of CO_2 fixation have been found in green plants—the Calvin cycle (or C_3 pathway) and the C_4 pathway. Which one of the following statements concerning these is *not* true?

a The C_4 pathway is more common in tropical plants than in temperate-zone plants.

b The C_3 pathway is more common in temperate-zone plants than in tropical plants.

c C_4 plants are more efficient than C_3 plants at fixing CO_2 when the concentration of available CO_2 is low.

d The C_4 pathway requires two different light-driven events, whereas the C_3 pathway requires only one.

e Some C_4 plants are now very important crop plants in the Middle West.

8 Which of the following does *not* occur in both photosynthesis and respiration?

a electron flow
b splitting of water molecules
c synthesis of ATP
d transfer of electrons to carrier molecules

9 Which one of the following statements best describes the relationship between photosynthesis and respiration?

a Respiration is the exact reversal of the biochemical pathways of photosynthesis.

b Photosynthesis stores energy in complex organic molecules, and respiration releases it.

c Photosynthesis takes place only in the light, and respiration takes place only in the dark.

d Photosynthesis occurs only in plants and respiration only in animals.

e ATP molecules are produced in photosynthesis and used up in respiration.

10 Of the total ATP yield possible from the complete oxidation of one molecule of glucose, approximately what percent is actually realized when a muscle must depend entirely on anaerobic metabolism?

a 5.5 percent   d 89 percent
b 11 percent   e 94 percent
c 17 percent

11 Cyanide blocks the respiratory transport system. As a result

a the Krebs cycle speeds up.

b electrons and hydrogens cannot flow from from NAD_{re} to oxygen.

c three ATPs are produced for every pair of electrons.

d production of water increases.

12 Which one of the following summary reactions (each describing a part of the process of metabolic breakdown of glucose) yields the most new ATP molecules, directly or indirectly, under aerobic conditions? (Disregard the fact that some equations are not fully balanced.)

a 2 pyruvic acid $\longrightarrow$ 2 acetyl-CoA

b glucose $\longrightarrow$ 2 pyruvic acid

c 2 PGAL $\longrightarrow$ 2 pyruvic acid

d 2 citric acid $\longrightarrow$ carbon dioxide + water

e 2 acetyl-CoA $\longrightarrow$ 2 citric acid

13 When glucose is broken down to carbon dioxide and water during aerobic respiration, more than 50 percent of its energy is released as

a oxygen.   d ATP.
b carbon dioxide.   e NAD_{re}.
c heat.

14 When a molecule of glucose is completely broken down in a cell to water and carbon dioxide, some ATP molecules are formed by substrate-level phosphorylation and some by oxidative phosphorylation via the electron-transport system. What percentage of the total number of ATP molecules formed comes from the latter process?

a 94 percent   d 78 percent
b 89 percent   e 6 percent
c 83 percent

15 Which one of the following is *not* associated with cold-blooded animals?

a body temperature close to environmental temperature

b low metabolic rate

c insulating hair or feathers

d metabolic rate that varies with environmental temperatures

e sluggish behavior in cold temperature

For further thought

1. How does ATP store and release energy? What do we do with all our ATP?

2. Compare the energy levels of glucose, pyruvic acid, NAD_{re}, ATP, CO_2, and H_2O.

3. Give an account of the formation of ATP, referring to each of the following molecules: H_2O, CO_2, O_2, 6-carbon carbohydrate, 3-carbon carbohydrate, 2-carbon molecule, 4-carbon molecule, 5-carbon molecule, 6-carbon molecule.

4. Describe the principal events in both cyclic and noncyclic photophosphorylation, and contrast these two processes.

5. Describe the Calvin cycle by which carbohydrates are synthesized from CO_2 in the dark reactions. Indicate the starting materials and the end products, and describe the relationship between photophosphorylation and carbon fixation.

6. Discuss structural features of the privet leaf particularly important in making it an efficient organ for carrying out photosynthesis.

7. Why must glycolysis take place in all cells?

8. Summarize the complete respiratory breakdown of one molecule of glucose, explaining (*a*) the main stages in the process, (*b*) the role of oxygen in the process, (*c*) the significance of the electron-carrier system, and (*d*) the fate of the energy contained in the glucose.

9. What is the relative efficiency of glycolysis and cellular respiration?

10. Why can oxygen be called a garbage truck for electrons?

11. How is it possible to use fats and proteins for energy sources?

Chapter 5

NUTRIENT PROCUREMENT AND PROCESSING

KEY CONCEPTS

1 In order to carry out their life processes, all organisms require prefabricated high-energy organic compounds or the raw materials from which they can be synthesized.

2 Autotrophic organisms manufacture their own organic compounds from inorganic raw materials absorbed directly from the environment.

3 Heterotrophic organisms must obtain high-energy organic compounds already synthesized; ultimately heterotrophs depend on nutrients synthesized by the autotrophs.

4 Much of the diversity among living things is a result of adaptations toward one of the three major nutritive modes: photosynthetic, absorptive, and ingestive.

5 Heterotrophs must digest their food into smaller molecules before their cells can absorb it.

6 Extracellular digestion is an adaptation for eating larger pieces of food than would be possible using only intracellular digestion; it is the general rule in multicellular animals.

7 A complete digestive tract has sections that are specialized for different functions; this produces an efficient digestive system.

OBJECTIVES

After studying this chapter and reflecting on it, you should be able to

1 Differentiate between the nutrient requirements of autotrophic and heterotrophic organisms.

2 State the nutrient requirements of the green plant and explain how the plant is adapted to procure these nutrients.

3. Explain why most of the mass of a plant's body comes from air, not from the solid earth in which it grows.

4. State the role of trace elements in both plants and animals.

5. Explain how water can move through the cortex of the plant root by flowing through the symplast or apoplast.

6. Draw several endodermal cells with their Casparian strips and relate their structure to the role of the endodermis in water and mineral absorption.

7. List the nutrients required by heterotrophs and give the function of each nutrient in the body.

8. Discuss the role of the various vitamins in the human body and indicate why a vitamin deficiency may lead to symptoms of disease.

9. Define an **essential amino acid** and explain why it is important to eat several different types of protein at each meal.

10. Differentiate between absorptive and ingestive heterotrophs and give an example of each.

11. Define these terms: saprophyte, parasite, herbivore, carnivore, omnivore.

12. Define digestion and explain why digestion of complex foods is necessary.

13. Differentiate between extracellular and intracellular digestion. Give the advantages of extracellular over intracellular digestion.

14. Explain what is meant by a "complete digestive tract" and the advantages of such a system.

15. Discuss the importance of mechanical breakup of bulk food in animals and give three examples of adaptations for this process.

16. Explain the adaptive significance of a storage chamber in the digestive tract.

17. Trace the route food follows from the time it enters your mouth until the indigestible residue leaves the anus. Indicate where most digestion and absorption occur.

18. Discuss the chemical changes that occur in food as it passes through your digestive tract. Give the function(s) of saliva, gastric juice, bile, pancreatic juice, and secretions of intestinal glands.

19. Explain how the lining of the small intestine is adapted to increase the absorptive surface area.

20. Give three functions of your large intestine.

21. Explain how organisms such as cows, rabbits, and horses can derive some nutrients from the cellulose they eat, even though mammals cannot digest cellulose.

SUMMARY

All organisms require high-energy organic compounds to carry out their life functions. Organisms capable of manufacturing their own organic nutrients from inorganic raw materials are *autotrophic*; those that cannot synthesize organic materials from inorganic raw materials and require prefabricated complex organic materials from the environment are *heterotrophic*. These two groups differ in both their nutrient requirements and the problems associated with nutrient procurement.

Nutrient procurement by green plants Most autotrophs are *photosynthetic*; they use light energy to drive their synthesis of organic compounds. Green plants, the largest group of photosynthetic organisms, require carbon dioxide and water as raw materials. In addition, they need certain mineral nutrients. Some minerals, required in substantial amounts (*macronutrients*), function as structural components of complex organic molecules; others, required in minute amounts (*micronutrients*), function as parts of enzymes or coenzymes.

The CO_2 required by the plant is absorbed by diffusion into cells in the interior of the leaf. Tiny openings or *stomata* in the epidermis allow the CO_2 to penetrate to the interior of the leaf, where it can circulate in the intercellular spaces and be available to individual cells.

Most higher land plants take in water and mineral nutrients from the soil through their roots. To absorb enough of these needed materials, plants require a sufficiently large surface area. As an organism gets bigger, its volume increases much faster than its surface area. Therefore, a large multicellular plant must have an enormous absorptive surface area to support its large volume. A typical root is extensively branched and subdivided and has many *root hairs*, tiny hairlike extensions of the epidermal cells, which vastly increase the total absorptive surface. Roots also store organic compounds and anchor the plant to the substrate.

For water and dissolved minerals to be available to other parts of the plant, they must first move through the outer tissues of the root into the *stele,* where the vascular tissues, which transport the nutrients, are located. The outer surface of the root is covered by *epidermis* that functions in water absorption. Inside the epidermis is an area of loosely packed parenchyma cells called the *cortex*. Water from the soil may be absorbed directly into the epidermal cells and move from cell to cell through the cortex to the stele by flowing through the interconnected cytoplasmic network of the cells, the *symplast*. Alternatively, it may move across the epidermis and cortex without entering a cell by flowing along the hydrophilic cell walls and intercellular spaces that make up the *apoplast*. However, the *endodermis*, which separates the cortex from the vascular cylinder, presents a barrier to such movement. The side and end walls of the endodermal cells contain a waxy, hydrophobic band (the *Casparian strip*) that forms a waterproof seal between the endodermal cells. Consequently, all materials must pass through the living endodermal cells to reach the stele, and the plant can exercise some control over the movement of substances dissolved in water into the vascular tissue.

Minerals are absorbed in ionic form from the soil water. The availability of inorganic ions for absorption depends on the fertility, physical structure, and pH of the soil. Each mineral is absorbed at its own rate, independent of the rate of absorption of the other minerals and also independent of the absorption of water. Simple diffusion, facilitated diffusion, and active transport are all involved in the absorption process.

A few photosynthetic plants supplement their inorganic diet with organic compounds obtained by trapping and digesting insects and other small animals. The nitrogenous compounds appear to be most beneficial to the plant.

Nutrients required by heterotrophs Since many of the organic molecules found in nature are too large to be absorbed unaltered through cell membranes, they must first be hydrolyzed (i.e. digested) by enzymes into their constituent building-block molecules.

There are three main groups of heterotrophic organisms: the bacteria, fungi, and animals. Bacteria and fungi are *absorptive heterotrophs*; they lack internal digestive systems and depend mainly on absorption as their mode of feeding. In contrast, animals are *ingestive heterotrophs*; they take in particulate or bulk food and digest it within chambers inside their body.

Many bacteria and fungi thrive on a diet containing only carbohydrates because they can synthesize other organic compounds from carbohydrates. However, most heterotrophic organisms require carbohydrates, fats, and proteins in bulk.

Nine *amino acids* are *essential* in the diet of most animals because the animals cannot synthesize them. Since amino acids cannot be stored in the body, all the essential amino acids must be ingested simultaneously and in the correct proportions if effective protein synthesis is to take place.

Some animals require no fat in their diet; they can interconvert carbohydrate and fat. Others cannot synthesize enough of certain fatty acids for their needs, so these *essential fatty acids* must be included in their diet.

Vitamins are organic compounds necessary in small quantities to given organisms that cannot synthesize them. Most vitamins function as coenzymes or parts of coenzymes. A prolonged vitamin deficiency impairs metabolic processes within the cell, often producing symptoms of a deficiency disease in the organism. Vitamins are classified into two groups on the basis of solubility; those of the B complex and vitamin C are water soluble, while vitamins A, D, E, and K are fat soluble.

Heterotrophs, like autotrophs, also require certain minerals, which are usually absorbed as ions.

Nutrient procurement by heterotrophs The fungi constitute a large, diverse group of sedentary heterotrophic organisms that live on or in their food supply. Their digestion is *extracellular*; digestive enzymes are secreted directly into the food supply and the products of digestion are then absorbed. Many fungi are *saprophytic* (living on dead material) while others are *parasitic* (living on or in other living organisms). A few fungi supplement their diets by trapping and digesting small animals.

Like fungi, animals must digest their food before it can cross their cell membranes. In some animals, digestion is *intracellular*; the food is ingested directly into the cell by endocytosis, then hydrolyzed in a food vacuole. In others, digestion takes place outside the cell (*extracellular digestion*), either in the environment or in a specialized digestive structure. In both types, digestion of complex food always precedes the absorption of substances across a membrane.

The unicellular protozoans carry on only intracellular digestion. A *food vacuole* is formed and fuses with a *lysosome* containing digestive enzymes. Digestion takes place within the vacuole, and the products of digestion are absorbed. Indigestible substances are expelled by exocytosis.

With the evolution of multicellularity came a corresponding evolution of cellular specialization resulting in a division of labor among cells. This evolutionary trend is clearly illustrated in the simple multicellular animals. The radially symmetrical coelenterates have a saclike body consisting of two principal cell layers surrounding a central digestive cavity with only one opening that must serve as both mouth and anus; this is a *gastrovascular cavity*. The "mouth" is surrounded by movable tentacles with numerous stinging structures called *nematocysts*, which are used in capturing prey. The specialized cells lining the cavity, the *gastrodermis*, carry on both extracellular and intracellular digestion. The products of digestion are distributed to cells specialized for other functions. Extracellular digestion allows the organism to eat larger pieces of food than it could handle using only intracellular digestion; it is the general rule in multicellular animals.

The free-living flatworms are bilaterally symmetrical, elongate animals with a gastrovascular cavity. They carry out some extracellular digestion, but most of the food particles are engulfed by the cells lining the gastrovascular cavity and digested intracellularly.

Animals above the level of coelenterates and flatworms have a *complete digestive tract*, i.e. one with two openings, a *mouth* and an *anus*. Food can now be passed in one direction through a tubular system that can be divided into a series of distinct sections, each specialized for a different function—mechanical breakup of bulk food, temporary storage, enzymatic digestion, absorption of products, reabsorption of water, storage of wastes, etc. The result is a more efficient digestive system.

The earthworm's digestive tract is a good example of this compartmental specialization.

The earthworm has a muscular *pharynx* for sucking in food, a *crop* for temporary storage, a *gizzard* for mechanical breakup, and an *intestine* with a large surface for extracellular enzymatic digestion and absorption.

Mechanical breakup of bulk food is common among animals, and a variety of structures that serve this function have evolved. However, not all animals eat large pieces of food and have special masticating devices; some animals, such as clams and mosquito larva, are filter feeders, straining small particles of organic matter from water.

A food storage chamber enables an animal to take in large amounts of food in a short time, and to utilize this food over a period of time. Such discontinuous feeding allows the animal to spend time searching for new food supplies and frees it for nonfeeding activities.

The digestive tract of humans consists of a series of chambers specialized for different functions. The first chamber is the *oral cavity*, where the teeth break up the food mechanically. The teeth of different vertebrates are specialized in a variety of ways and may differ in number, structure, and arrangement. Sharp pointed teeth, poorly adapted for chewing, characterize *carnivores* (meat eaters) whereas broad flat teeth, well adapted for chewing, characterize *herbivores* (animals that eat plant materials). Humans, who eat both plant and animal material, are *omnivores*; their teeth are rather unspecialized.

The muscular tongue manipulates the food during chewing and mixes it with *saliva*. The lubricated food is pushed backward through the *pharynx* into the *esophagus*. The food moves quickly through the long esophagus, pushed along by waves of muscular contraction called *peristalsis*, into the *stomach*, where it is stored and further broken up by the squeezing and churning action of the stomach muscles. Pepsin secreted in the *gastric juice* begins the process of protein digestion. The food moves from the stomach into the first part of the small intestine, the *duodenum*, where ducts carrying secretions from the liver and pancreas enter. The rest of the *small intestine* is long and coiled. Most of the digestion and the absorption of the products of digestion occurs in the small intestine.

The length of the small intestine in various animal species is proportional to the amount of plant material in their diet; herbivores ordinarily have a longer small intestine than carnivores, because the cellulose cell walls of plants are difficult to break up and tend to interfere with digestion and absorption. Several structural adaptations other than length increase the absorptive surface area. The *mucosa* that lines the intestine is thrown into numerous folds from which small fingerlike outgrowths, called *villi*, protrude. In addition, the epithelial cells covering the villi have numerous *microvilli*, closely packed, cylindrical processes, on their surface.

Indigestible material, water, and unabsorbed substances move on into the *large intestine* or *colon*. A blind sac, or *caecum*, projects from the junction of the small and large intestine. In some mammals the caecum is large and contains microorganisms capable of digesting cellulose. Since mammals cannot digest cellulose themselves, this microbial action is beneficial. Two important functions of the large intestine are the reabsorption of water and the excretion of certain salts when their concentration in the blood is too high. The last portion of the large intestine, the *rectum*, functions as a storage chamber for the feces until defecation.

Enzymatic digestion, the complete hydrolysis of organic nutrients into their building-block molecules, occurs mainly in the mouth, stomach, and small intestine. It is generally a two-step process. First the complex nutrients are hydrolyzed enzymatically into smaller fragments, then other enzymes complete the hydrolysis to the building-block compounds. For example, *amylase* in saliva and pancreatic juice hydrolyzes starch to double sugars, and intestinal enzymes split the double sugars into simple sugars. In protein digestion, the large polypeptide chains are first hydrolyzed into smaller fragments by *endopeptidases*, and the digestive process is completed by the highly specific *exopeptidases*, which split off the terminal amino acids. Absorption of the simple sugars and amino acids into the blood involves active transport. A summary of the action of the various enzymes (with their sites of production shown in parentheses) is shown below.

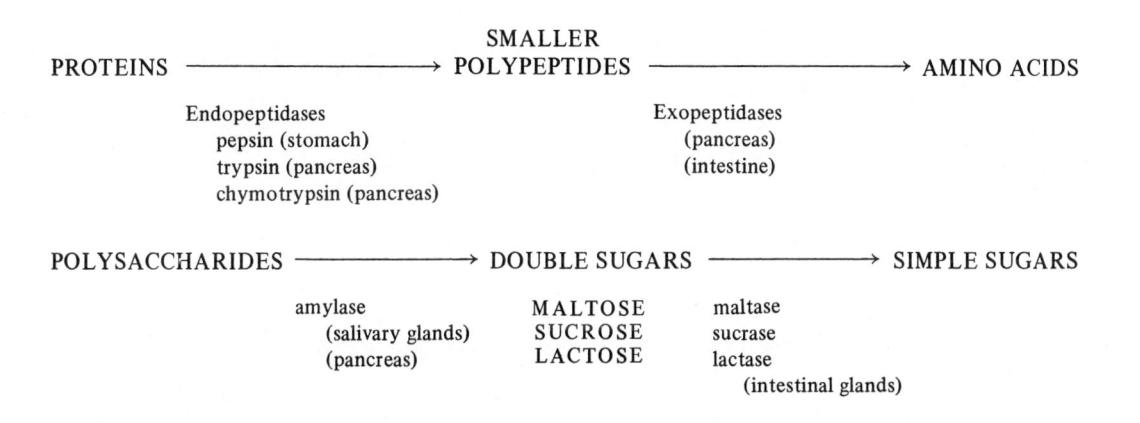

PROTEINS ⟶ SMALLER POLYPEPTIDES ⟶ AMINO ACIDS

    Endopeptidases   Exopeptidases
      pepsin (stomach)   (pancreas)
      trypsin (pancreas)   (intestine)
      chymotrypsin (pancreas)

POLYSACCHARIDES ⟶ DOUBLE SUGARS ⟶ SIMPLE SUGARS

          amylase   MALTOSE   maltase
        (salivary glands)   SUCROSE   sucrase
        (pancreas)   LACTOSE   lactase
                                                  (intestinal glands)

The digestion and absorption of fat follows a different pattern. Much fat is absorbed directly since it is lipid-soluble and can cross the cell membrane. Some fat is digested; the liver produces *bile* that acts to emulsify the fats so the principal fat-digesting enzyme, *lipase*, can hydrolyze them more effectively.

         emulsification

FATS ⟶ EMULSIFIED FAT ⟶ FATTY ACIDS AND GLYCEROL

    bile salts (liver)   lipase (pancreas)

QUESTIONS

Testing recall

Referring to the cross section of a root, match each item below with the associated part of the root.

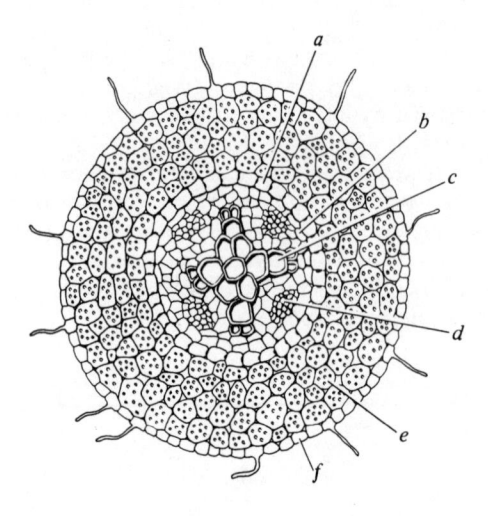

1 region that functions in carbohydrate storage
2 outer boundary of the stele
3 area where much water absorption occurs through its enormous surface area
4 cells with a waterproof band along side and end walls
5 structure that conducts water and dissolved minerals to other parts of the plant body
6 area where water and dissolved minerals must cross a cell membrane to reach the vascular tissue
7 area that contains meristematic cells that may give rise to lateral roots
8 region where adjacent plant cells are interconnected, forming a symplast through which water can flow

Match the items listed below with the associated parts of the digestive system in the drawing.

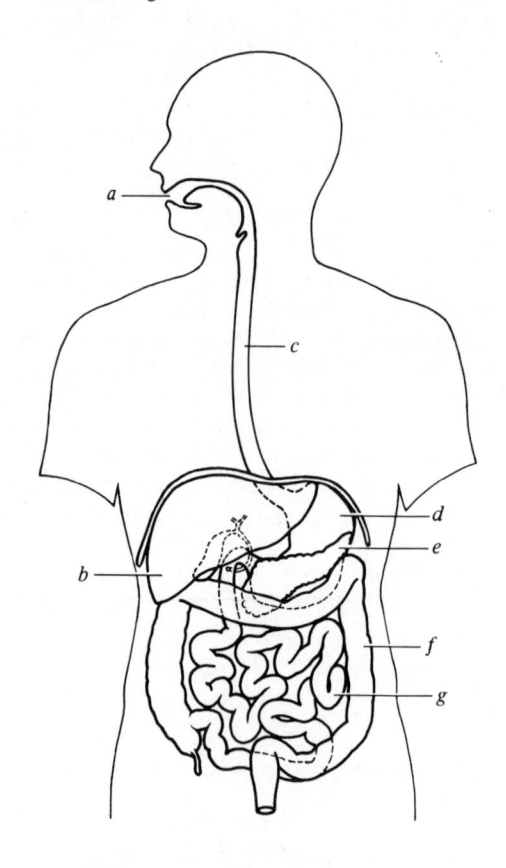

9 Carbohydrate digestion begins here.
10 Protein digestion begins here.
11 Enzymes that digest fat, protein, and carbohydrates are produced here.
12 Most water reabsorption occurs here.
13 Most fat digestion occurs here.
14 Most products of digestion are absorbed here.
15 Fat-digesting enzymes are synthesized here.
16 This organ produces a substance that emulsifies fat.
17 This organ secretes enzymes active at a low pH.
18 This structure is lined by villi.
19 In cattle, most cellulose digestion takes place in a highly modified form of this structure.

Testing knowledge and understanding

Choose the one best answer.

1 Autotrophic organisms
   a must digest their nutrients before taking them into the cell.
   b synthesize their own organic materials from inorganic materials in the environment.
   c require complex organic molecules already synthesized by other organisms.
   d require no external energy source since they synthesize their own high-energy compounds.

2 Three macronutrients needed by green plants and included in considerable quantity in most fertilizers are
   a calcium, boron, and lead.
   b carbon dioxide, water, and nitrogen.
   c copper, zinc, and sodium.
   d glucose, PGAL, and sucrose.
   e nitrogen, phosphorus, and potassium.

3 Nitrogen is needed for the formation of
   a sugars.
   b fats.
   c cellulose.
   d proteins.
   e starches.

4 Of the four most abundant elements in most plants (C, H, O, and N), which does a terrestrial green plant procure mainly through its roots from the soil?
   a H and O
   b C and O
   c O and N
   d H and N
   e C and N

5 The epithelial cells lining your digestive tract effectively control which substances will be absorbed and transported to other parts of your body. Because their walls have an affinity for water, the epidermal cells of roots are much less efficient in this sort of regulation. Nevertheless, plants do rely on membrane selectivity to control which substances will enter the xylem for transport. The cell layer most responsible for this regulation is the
   a cortex.
   b endodermis.
   c pericycle.
   d xylem.
   e pith.

6 Which one of the following statements most accurately describes an essential amino acid?
   a It can be obtained in the diet only from animal products such as meat, eggs, milk, and cheese.
   b It is a universal building block for the body's proteins, found in all the proteins our bodies synthesize.
   c Although it is necessary for life, the organism in question cannot synthesize it and hence must obtain it in the diet.
   d It functions as a coenzyme in a biochemical pathway that is essential for life.
   e It is essential for all animals but can be synthesized only by green plants.

7 A deficiency of vitamin C (ascorbic acid) in the human diet results in the disease called scurvy, characterized by bleeding gums, loosening of teeth, delayed healing of wounds, and painful and swollen joints. These deficiency symptoms are understandable because ascorbic acid plays a major role in
   a forming red blood cells.
   b maintaining good night vision.
   c facilitating calcium absorption.
   d promoting blood clotting.
   e forming collagen fibers in connective tissue.

8 Beriberi can be prevented or cured if the diet contains sufficient
   a vitamin A.
   b vitamin B_1.
   c vitamin C.
   d vitamin D.
   e vitamin K.

9 Which one of the following carries on *only* intracellular digestion?
   a human
   b hydra
   c mushroom
   d *Paramecium*
   e earthworm

10 Nutritionally, a lion could best be described as an
   a autotroph.
   b absorptive heterotroph.
   c ingestive heterotroph.
   d herbivore.
   e omnivore.

11 Gastrovascular cavities
   a are found only in unicellular organisms such as *Paramecium.*
   b have only one opening to the exterior, which functions as both mouth and anus.
   c generally include a gizzard instead of teeth to break up bulk food.
   d are commonly found in herbivorous animals such as sheep and goats.
   e usually have several chambers, each specialized for a different function.

12 Which one of the following animals has the least subdivision of its digestive system into separate chambers specialized for different operations on food?

 a earthworm   d hydra
 b shark   e chicken
 c mosquito

13 Which one of these animals would you expect to have the most developed grinding surface on the molar teeth?

 a dog   d sheep
 b cat   e lion
 c human

14 Which one of the following is *not* true of the digestive system of most higher land animals?

 a It begins with a mouth and ends with an anus.
 b It frequently depends on microorganisms to perform vital functions.
 c The surface area is reduced as much as possible to permit easy passage of food.
 d It has a mechanism for chewing and grinding the food.
 e It includes a chamber for food storage.

15 Which one of the following digestive enzymes would be most likely to catalyze the reaction

$$C_{12}H_{22}O_{11} + H_2O \longrightarrow C_6H_{12}O_6 + C_6H_{12}O_6$$

 a amylase   d endopeptidase
 b sucrase   e exopeptidase
 c lipase

16 Which one of the following mammals would you expect to have the longest small intestine relative to its body size?

 a human   d seal
 b dog   e rabbit
 c cat

17 A hungry student eats a ham and cheese sandwich and drinks a glass of milk for lunch. He has a piece of apple pie for dessert. Digestion of the starch in this meal is carried out in the

 a oral cavity and stomach.
 b stomach and small intestine.
 c small intestine and large intestine.
 d oral cavity and small intestine.
 e stomach and large intestine.

18 The products of digestion of the protein in the student's meal are absorbed in the

 a stomach.   d pancreas.
 b small intestine. e large intestine.
 c liver.

19 Enzymatic hydrolysis in the mouth will primarily affect the

 a ham.   d milk sugar.
 b cheese.   e milk fats.
 c bread.

20 The digestion of fats takes place almost entirely in the

 a mouth.   d small intestine.
 b esophagus.   e large intestine.
 c stomach.

21 The enzyme that catalyzes hydrolysis of fat is

 a amylase.   d trypsin.
 b pepsin.   e sucrase.
 c lipase.

22 Bile aids in fat digestion by

 a hydrolyzing the bonds between glycerol and fatty acids.
 b breaking peptide bonds.
 c converting unsaturated fats to saturated fats.
 d emulsifying fat droplets so that more surface area is exposed to the action of digestive enzymes.
 e stimulating increased release of fat-digesting enzymes from the pancreas.

For further thought

1 Contrast the nutrient requirements of animals and green plants. Why are animals ultimately dependent on green plants?

2 Describe the processes involved in the uptake of water and minerals by a plant root.

3 Define "vitamin" and make a general statement of how vitamins function. Choose one vitamin to discuss, including characteristic symptoms of its deficiency, foods in which it is found, etc.

4 What is the general function of the digestive enzymes? Name some factors in the digestive tract that influence the function of the enzymes and discuss their effect. Describe some specific ways enzymes may "cut up" a protein molecule.

5 Discuss and compare adaptations for nutrient procurement and processing in hydra, earthworm, cow, and human being.

6 Follow a hamburger with roll through the digestive tract. Name the important enzymes involved, their source, and their actions. What happens to the end products?

Chapter 6

GAS EXCHANGE

KEY CONCEPTS

1 A basic problem for most living organisms, both plant and animal, is procuring oxygen for respiration and eliminating carbon dioxide.

2 Gas exchange between a living cell and its environment always takes place by diffusion across a thin, moist cell membrane.

3 All organisms require a protected respiratory surface of adequate dimensions relative to their volume. The larger the organism, the more surface it requires.

4 Many organisms need a transport system to carry the gases from the respiratory surface to the internal cells.

OBJECTIVES

After studying this chapter and reflecting on it, you should be able to

1 Give the reason why most organisms require molecular oxygen.

2 Explain why plants require a supply of oxygen even though they produce it during photosynthesis.

3 Give a reason why the cell membrane must be moist for gas exchange to occur.

4 Explain why an organism such as a goldfish has evolved a special respiratory surface whereas the large algae have no special gas exchange mechanism.

5 Distinguish between an invaginated and an evaginated respiratory surface. Give an example of each.

6 State the four basic requirements for a respiratory system and indicate how the leaves, roots, and stems of a typical terrestrial plant are adapted to meet these requirements.

7 Give three reasons why obtaining oxygen is a greater problem for aquatic organisms than for air breathers.

8 Explain how a segmented marine worm and a fish meet their gas exchange requirements.

9 Give three reasons why most land animals have evolved invaginated rather than evaginated respiratory systems.

10 Trace the route a molecule of air follows in your body from inhalation to your lung.

11 Explain how the structure of your lung satisfies the four basic requirements for a respiratory system.

12 Describe the process of breathing in your body, and explain how this differs from the breathing process in a frog.

13 Give reasons why birds and mammals have such high oxygen requirements, and why birds obtain oxygen from the air more efficiently than mammals.

14 Outline the evolution of the swim bladder of modern fish, and explain how the swim bladder allows a fish to swim at different depths.

15 Describe the tracheal system of an insect and identify two fundamental ways in which this invaginated system differs from that of the land vertebrates.

SUMMARY

Since aerobic respiration is the chief method of respiration in both plants and animals, the vast majority of organisms must have some method of obtaining oxygen and getting rid of waste carbon dioxide.

Gas exchange between a living cell and its environment always takes place by *diffusion* across a moist cell membrane. The cells of most small aquatic organisms are in close contact with the surrounding medium and carry out gas exchange across their whole body surface; no special respiratory devices are necessary. With the evolution of large three-dimensional organisms came the necessity for a corresponding evolution of specialized respiratory surfaces to meet the increased oxygen requirements. A wide variety of structures for gas exchange have evolved, but each respiratory system must meet four basic requirements:

1 A respiratory surface of adequate dimensions
2 A moist exchange surface
3 A method of transporting the gases to the cells
4 Protection for the fragile respiratory surface

Gas exchange in plants Gas exchange associated with photosynthesis and cell respiration takes place at a high rate in the leaves. The loosely packed mesophyll cells in the interior of the leaf provide an adequate surface area for gas exchange. The epidermis with its waxy cuticle protects the mesophyll from mechanical damage and water loss. Tiny openings in the epidermis allow the gases to penetrate to the interior, where they circulate through the numerous intercellular spaces and reach each cell. The exchange surfaces remain moist because they are exposed to air only in the intercellular spaces, where the humidity is nearly 100 percent.

The size of each opening, or *stoma*, in the epidermis is regulated by two bean-shaped *guard cells*. When the guard cells are turgid (usually in the light), the stoma is open; when the guard cells lose water (usually in the dark), the stoma is closed. When the stomata are open, gases can move into the interior of the leaf, but at the same time the plant loses water vapor in the process called *transpiration*. The requirements of water conservation conflict with those of photosynthesis and respiration. If photosynthesis is to occur at a high rate, the plant must obtain sufficient carbon dioxide; hence, the stomata must be open. Excessive water loss through transpiration is prevented by the vascular system, which replaces water at a steady rate. If water loss is too rapid, the plant may wilt; when the guard cells lose water they close the stomata, preventing further water loss.

We still do not know how the plant regulates the size of the stomatal openings through the rapid turgidity changes in the guard cells, but unusual starch metabolism, pH changes, and K^+ levels in the guard cells appear to be involved.

Root cells obtain oxygen by simple diffusion across the moist membranes of the individual cells; air reaches the interior cells via the intercellular spaces. Gas exchange in stems takes place through areas of loosely arranged cells called *lenticels* connected with an extensive system of air-filled intercellular spaces that penetrates to all parts of the plant body.

Gas exchange in animals Most large multicellular animals have evolved some type of special respiratory surface. These may be grouped in two categories: outward-oriented (*evaginated*) and inward-oriented (*invaginated*) extensions of the body surface.

Gas exchange in aquatic animals Most multicellular aquatic animals utilize evaginated exchange surfaces called *gills*. Most gills have such finely subdivided surfaces that a few small gills may expose an immense total exchange surface to the water. Since most gills contain a rich supply of blood vessels, oxygen can move by diffusion from the water, across the intervening cells, and into the blood, where special carrier pigments transport it to the individual cells throughout the body. Carbon dioxide moves in the opposite direction.

In fish, the water flows over the gills in the opposite direction to that of the blood flow; this *countercurrent exchange system* maximizes the amount of O_2 the blood can pick up from the water.

Because of oxygen's low solubility and slow diffusion rate in water, most aquatic organisms must constantly move water across the exchange surface.

Gas exchange in terrestrial animals Most terrestrial animals have evolved invaginated respiratory systems, of which the two principal types are *lungs* and *tracheae*. Lungs, invaginated gas-exchange organs supplied with blood vessels, are typical of land snails and the higher vertebrates. It is believed that the ancestral fish from which land vertebrates evolved had primitive lungs. The evolution of the lung in higher vertebrates has tended toward increased surface area (through folding and subdivision of the inner surface) and an increased blood supply to the exchange surface.

In human beings, air is drawn through the *external nares* into the *nasal cavities*, which function in warming, moistening, and filtering the air, and in smelling. Next air moves into the throat or *pharynx*, a common passageway for food and air; the air and food passages cross in the pharynx. After leaving the pharynx through a ventral opening, the *glottis*, the air enters the *larynx*, or voice box. During swallowing, the larynx is raised and presses against a small flap of tissue, the *epiglottis*, closing the glottis.

The *trachea* is a noncollapsible air duct leading from the larynx to the *thoracic cavity*. At its lower end the trachea divides into the two *bronchi*, which lead into the two lungs. Each bronchus branches and rebranches, and the *bronchioles* thus formed branch repeatedly into smaller ducts that ultimately terminate in grapelike clusters of air sacs called *alveoli*. Each alveolus is surrounded by a dense bed of blood capillaries; the diffusion of O_2 and CO_2 takes place here.

Air is drawn into and pushed out of the lungs by the mechanical process called *breathing*. When air is forced into the lungs the process is termed *positive-pressure breathing*; when it is drawn into the lungs as a result of an enlarged body cavity it is termed *negative-pressure breathing*. Birds and mammals utilize negative-pressure breathing. In humans, inhalation occurs when the rib muscles contract, drawing the rib cage up and out, and the *diaphragm* contracts, moving downward. These two actions increase the volume of the thoracic cavity, thus reducing the air pressure within the cavity below the atmospheric pressure and drawing air into the lungs. Normal exhalation is a passive process; the muscles relax and the rib cage and diaphragm return to their resting position, thus decreasing the thoracic volume and increasing the air pressure in the lungs, which forces air out.

The pattern of air flow in the respiratory system of birds is fundamentally different from that of mammals. The special arrangement of the lungs and their associated *air sacs* permits a continuous unidirectional flow of air through the lungs during both inhalation and exhalation. Because the blood flow in the capillaries is at an angle to the continuous air flow, the blood can pick up the maximum amount of O_2; thus birds extract oxygen from air more efficiently than mammals.

Modern fish do not have lungs as the ancestral fish did; it is believed that the primitive lung evolved into the *swim bladder* of modern fish. The swim bladder is filled with gases and acts to adjust the density of the fish to that of the water at different depths.

The tracheal system typical of land arthropods consists of many small air ducts called *tracheae* that run from openings (*spiracles*) in the body wall and carry air directly to the individual cells. There is no significant transport of O_2 by the blood.

QUESTIONS

Testing recall

Complete the following statements.

1 To obtain enough energy to carry on life processes, most plants and animals require _____ for respiration.

2 Aquatic organisms such as coelenterates have no special respiratory mechanism since gases can move a distance of approximately _____ mm by simple diffusion.

3 Gas exchange between cells and the environment always takes place by _____ across _____ membranes.

4 Gas exchange in the leaf takes place across the cell membranes of the _____ cells.

5 Stomata are usually _____ when the guard cells are carrying out photosynthesis in bright sunlight.

6 In the leaf, excessive water loss by transpiration is prevented by _____.

7 Gaseous oxygen can reach living cells deep within the plant stem by entering the intercellular air space system via openings called _____.

8 Gills are an example of an _____ respiratory surface whereas lungs are an example of an _____ surface.

9 In the gills of fish, the countercurrent flow of water and blood greatly _____ the pickup of O_2 by the blood.

10 The rate of diffusion is _____ in air than in water.

11 Most land organisms utilize an _____ type of respiratory system of which there are two principal types, _____ and _____.

12 Gas exchange in the human takes place by the process of _____ across the thin, moist membranes of the _____, which are surrounded by a dense network of _____.

13 Incoming air is warmed, moistened, and filtered in the _____.

14 Frogs force air into their lungs in a process called _____-pressure breathing whereas birds and mammals _____ air into their lungs using _____-pressure breathing.

15 When the diaphragm contracts, the air pressure in the lungs _____.

16 Swim bladders probably evolved from the _____ of primitive fish.

17 _____ are the respiratory openings on the body surface of insects. They lead into many smaller air ducts called _____, which conduct air to internal cells.

18 List the four basic requirements for a specialized respiratory surface.

Testing knowledge and understanding

Choose the one best answer.

1 The principal site of gas exchange in the leaf is the

a upper epidermis.   *d* mesophyll.
b cuticle.   *e* lower epidermis.
c lenticel.

2 During the day, the guard cells

a lose water and are flaccid.
b synthesize starch from sugar.
c have a relatively low concentration of K^+.
d have a lower osmotic concentration than at night.
e carry out photosynthesis.

3 Which of the following features do *all* respiratory systems have in common?

   a The exchange surfaces are moist.
   b They are enclosed in a special chamber.
   c They are maintained at a constant temperature.
   d They are exposed to air.
   e They are found only in vertebrates.

4 Animals like the earthworm must live in moist soil. How is this related to their respiration?

   a They respire through their moist skin.
   b Their lungs function most efficiently in a moist habitat.
   c Waterlogged soil is rich in oxygen.
   d The soil must be moist to keep their invaginated respiratory surfaces moist.

5 In which one of the following habitats would aquatic organisms be most likely to show special adaptations to low oxygen levels?

   a a pool at the base of a woodland waterfall
   b the Antarctic Ocean
   c a fast-running mountain stream
   d a shallow pool in a sunny meadow in Texas
   e the Great Lakes

6 Which of these statements is *not* true for gills operating in water?

   a Water can support the delicate gill filaments.
   b Water is difficult to move over the gill surfaces.
   c Keeping membranes moist is no problem.
   d Water carries more oxygen than air.
   e Oxygen content in water depends on temperature.

7 The normal air pathway in the human respiratory system is

   a trachea $\longrightarrow$ pharynx $\longrightarrow$ alveolus $\longrightarrow$ bronchus.
   b bronchus $\longrightarrow$ bronchiole $\longrightarrow$ pharynx $\longrightarrow$ trachea.
   c alveolus $\longrightarrow$ bronchus $\longrightarrow$ bronchiole $\longrightarrow$ pharynx.
   d trachea $\longrightarrow$ bronchus $\longrightarrow$ alveolus $\longrightarrow$ pharynx.
   e pharynx $\longrightarrow$ trachea $\longrightarrow$ bronchus $\longrightarrow$ alveolus.

8 Which one of these statements about human lungs is *not* true?

   a Gas exchange always takes place across moist membranes.
   b The gases move across the exchange membranes by simple diffusion.
   c The total exchange surface area is very large.
   d The walls of the alveoli are only one cell thick.
   e The partial pressure of CO_2 is higher in the air in the lungs than in the blood in the alveolar capillaries.

9 In the frog, which structure is *not* important in respiration?

   a nostrils   *d* diaphragm
   b mouth   *e* skin
   c lungs

10 In the respiratory system of birds, oxygen-rich air moves into the lungs

   a during inhalation.
   b during exhalation.
   c during both inhalation and exhalation.
   d only when the air sacs contract.
   e when air is forced from the mouth into the trachea.

11 Insects carry out respiratory gas exchange

   a in specialized external gills.
   b in specialized internal gills.
   c in the alveoli of their lungs.
   d across the membranes of nearly every internal cell.
   e across their thin moist skin.

For further thought

1 Explain why plants do not need special gas-transporting systems whereas many animals require some sort of transport mechanism to carry gases from the exchange surface to the rest of the body.

2 Discuss basic requirements for gas exchange mechanisms, and show how these are fulfilled in a fish, a human, an insect, and a large tree.

3 Why are gills not suitable for life on land?

4 The second largest invertebrate group is the phylum Mollusca. The name means "soft bodied"; the soft body is wholly or partially surrounded by a thin fleshy mantle and is commonly sheltered in an external limy shell. A familiar example of the mollusc is the marine clam. In the marine clam, gas exchange is performed by

two double gills, which hang in the mantle cavity. The gills consist of many interconnected thin plates or lamellae. Each lamella is supplied with small blood vessels, which bring blood to the gill and return it to the heart. Cilia cover the gills and function to produce a flow of water through the gills.

a For gas exchange, give one advantage and one disadvantage of a marine environment.
b Discuss how the clam meets the basic requirements of a gas exchange system.

5 Compare and contrast the respiratory systems of a human and a terrestrial insect.

6 Cigarette smoke has been found to have the following effects on the respiratory system:

a destruction of many of the cilia that line parts of the respiratory tract;
b thickening of the walls of the bronchioles, thus reducing the interior diameter of the tubes;
c rupturing of the walls of some of the alveoli.

For each of these effects, indicate how the *normal* functioning of the respiratory tract is altered by the use of cigarettes.

7 You are probably aware that your exhaled breath contains water vapor. How do you account for this?

Chapter 7

INTERNAL TRANSPORT

KEY CONCEPTS

1. Diffusion plays an important role in the movement of materials, but it is a very slow process. Consequently most organisms have evolved some sort of specialized transport mechanism to distribute substances more rapidly.

2. The successful exploitation of the land environment by plants was dependent on the evolution of specialized conducting tissues that would transport water and nutrients from one part of the plant to another.

3. The active way of life and rapid metabolism of animals require that each cell be provided with a continuous, abundant supply of oxygen and nutrients to metabolize for energy. Therefore many complex animals have evolved a true circulatory system to transport materials to and from the body cells more efficiently.

4. The high metabolic rate and active way of life of the warm-blooded birds and mammals are made possible by the four-chambered heart and the evolution of the separate pulmonary and systemic circulations.

5. Nearly all exchange of materials between the blood and the tissue fluid occurs in the capillaries; this exchange is governed by a delicate balance between the blood pressure and osmotic pressure.

6. The cells of higher animals require a fairly uniform, or stable, environment in order to maintain normal function. The circulatory system plays an important role in maintaining this stability within the body.

OBJECTIVES

After studying this chapter and reflecting on it, you should be able to

1. Discuss the role that diffusion and intracellular processes play in the movement of materials in unicellular and multicellular animals.

2. Explain why many very small animals can get along without a specialized circulatory system whereas most higher animals cannot.

3. Give three reasons why the evolution of specialized transport tissue was necessary for the higher plants to exploit the land environment fully.

4. Contrast the structure of the xylem tissue with that of the phloem with respect to the types of cells present, the structure of the conducting cells, and whether the conducting cells are living or dead.

5. Distinguish between primary and secondary tissue in a dicot and describe the process of secondary growth in a dicot stem. Show by diagrams how the structure of a typical 25-year-old tree trunk (as seen in cross section) would differ from that of a stem of the same species during its first season of growth.

6. Discuss and evaluate possible mechanisms for the rise of sap and the translocation of organic solutes in vascular plants.

7. Give two major ways in which the circulatory system of plants differs from that of animals.

8. Differentiate between an open and closed circulatory system and compare these in speed and efficiency. Explain why insects can function so successfully with an open circulatory system whereas animals such as earthworms utilize a closed system.

9. Describe the principal parts of your heart and the main blood vessels and review the general course of blood circulation. Explain how your heart differs from those of fish and amphibians, and the significance of these differences.

10. List in order the events that take place during the cardiac cycle.

11. Explain why the blood pressure in the capillaries and veins is considerably lower than that in the arteries and why the blood pressure is constant rather than fluctuating.

12. Compare the structure and function of the arteries, veins, and capillaries.

13. Describe the structure of the lymphatic system and discuss three important functions it performs to help maintain the normal functions of the body.

14. List four main components of the blood and give a major function of each.

15. Explain the process of blood clotting and show, by diagram, the main parts of the mechanism.

16. Explain how the structure of hemoglobin is related to its function and indicate the conditions under which oxyhemoglobin is formed and dissociated.

17. Explain how CO_2 is transported in the blood and the role hemoglobin plays in this process.

18. State two ways in which the blood acts as a protective agent against infection and disease.

SUMMARY

In general, only very small organisms can rely exclusively on the processes of diffusion and intracellular transport for the movement of substances; most higher organisms, both plant and animal, require a specialized transport system to deliver required materials to the cells and remove the waste products.

Transport in higher plants The large multicellular land plants, the *vascular plants*, have evolved two types of specialized conducting tissue: *xylem* and *phloem*. These tissues form continuous pathways running through the roots, stems, and leaves, and serve to transport materials from one part of the plant body to another.

Stems function as organs of transport and support. The arrangement of the vascular tissue varies in stems of different types and ages. In a young dicot stem the vascular tissue may be arranged in a continuous hollow cylinder around the central pith or in a ring of discrete bundles. The phloem always lies outside the xylem, with a layer of lateral meristematic tissue, the *vascular cambium*, between them. In monocot stems the vascular tissue is in discrete bundles scattered throughout the stem, and there is no vascular cambium. In herbaceous and young woody stems the tissues are composed of cells produced by the *apical meristem* as the stem grows in length; these tissues are *primary tissues*.

In some species of dicots the cambium becomes active and produces new cells; those formed on the outside differentiate as *secondary phloem*, those formed on the inside as *secondary xylem* or wood. All tissues derived from the cambium are *secondary tissues*; they contribute to growth in diameter. Since xylem cells produced early in the growing season are larger than cells produced later, a series of concentric annual rings is formed. Usually only the newer, outer rings, or *sapwood*, are functional in transport; the cells of the older rings, the *heartwood*, have become plugged.

As woody stems grow in diameter, a layer of cells outside the phloem, the *cork cambium*, begins to produce cork cells, which provide a waterproof coating for the plant.

The sequence of tissues (moving from outside toward the center) in an old woody stem is: cork tissue (outer bark), cork cambium, primary phloem, secondary phloem (inner bark), vascular cambium, secondary xylem (wood), primary xylem, and pith.

Xylem contains two types of conductive cells: *tracheids* and *vessel elements*. Both are dead at functional maturity; the cellular contents disintegrate, leaving thick, lignified cell walls that form hollow tubes within which vertical movement of materials can take place. Numerous *pits* in the wall allow the lateral movement of materials from cell to cell. Xylem also contains *fiber* and *parenchyma* cells. The thick-walled fiber cells are supportive elements; the parenchyma cells are often arranged to form *rays*, which function as pathways for the lateral movement of materials.

Phloem also contains both fiber and parenchyma cells in addition to those unique to it: *sieve elements* and *companion cells*. The elongate sieve elements with their perforated end walls—the *sieve plates*—are arranged end to end, forming a *sieve tube* through which vertical transport takes place. At maturity the nucleus of the sieve element disintegrates, but the cytoplasm remains. Companion cells, which retain both their nuclei and their cytoplasm, are intimately associated with the sieve elements in most advanced plants.

The ascent of sap Water and inorganic ions absorbed by the roots (sap) move upward through the plant body in the tracheids and vessels of the xylem. The most widely accepted theory for the ascent of sap is the *cohesion theory*. As water is lost by transpiration from the aerial parts of the plant, the water is replaced by withdrawal of water from the water column in the xylem. The water in the xylem forms a continuous chain from the aerial parts of the plant to the roots; these water molecules exhibit great cohesion because of hydrogen bonding. Thus the withdrawal of water molecules from the top of a water column will pull the whole chain of water molecules upward.

In some plants, especially in early spring, *root pressure* may contribute to the upward movement of sap in the xylem. Root pressure is created when ions are actively transported into the stele, and water follows passively in such quantity as to build up pressure in the xylem, pushing the sap slowly upward.

Translocation of solutes Inorganic ions are transported, or *translocated*, upward from the roots to the leaves primarily through the xylem; however, the downward transport of these ions is through the phloem. Most organic solutes (carbohydrates and organic nitrogen compounds) are translocated, both up and down, through the phloem.

Unlike the xylem transport, the movement of substances in the phloem is through living cells that retain their cytoplasm. The end walls are penetrated only by the tiny pores of the sieve plate. The mechanism by which this transport takes place is not well understood. The most widely supported hypothesis of phloem function is the *pressure flow* hypothesis, according to which there is a mass flow of water and solutes under pressure through the sieve tubes from an area of high turgor pressure (the "source") to an area of low turgor pressure (the "sink").

Circulation in higher animals Most higher animals have a true circulatory system in which blood is moved around the body along a fairly definite path. Animal circulatory systems usually have at least one pumping device, or *heart*. The system is a *closed circulatory system* if the blood is always contained within well-defined vessels; it is an *open circulatory system* if blood leaves the vessels and flows through large open spaces within the body. Closed circulatory systems are characteristic of earthworms and all vertebrates. Open circulatory systems are characteristic of most molluscs and all arthropods.

The open circulatory system of insects consists of a dorsal longitudinal vessel with valved openings at the posterior end that allow blood to move into the vessel. This vessel, or "heart," contracts, forcing the blood out at the anterior end. The blood then flows backward through large sinuses and finally reenters the dorsal vessel. Contraction of the muscles of the body wall and gut accelerates the movement of the blood.

The closed circulatory system characteristic of vertebrates consists of a heart and numerous vessels: *arteries*, which carry blood away from the heart; *veins*, which carry blood to the heart; and *capillaries*, tiny vessels that connect the arteries to the veins. The actual exchange of materials between the blood and tissues takes place in the capillaries.

Human circulatory system The human heart is a double pump; each side of the heart is divided into two chambers, an upper *atrium* that receives the blood and pumps it into the lower chamber, or *ventricle*, which then pumps the blood away from the heart.

The right heart receives deoxygenated blood from all over the body and pumps it via the *pulmonary arteries* to the lungs, where it picks up O_2 and gives up CO_2. The oxygenated blood then returns to the left atrium by the *pulmonary veins*. This portion of the circulatory system is called the *pulmonary circulation*.

The left ventricle pumps the blood into the *aorta* and its numerous branches, from which it moves into capillaries, where the exchange of materials takes place, then into veins, and finally back via the *anterior* or *posterior vena cava* to the right side of the heart. This portion of the circulatory system is called the *systemic circulation*.

The four-chambered heart, with complete separation of the right and left sides, is characteristic of the "warm-blooded" birds and mammals. The hearts of primitive vertebrates and modern fish have only one atrium and ventricle; the blood is pumped directly to the gills for oxygenation and on into the systemic circulation. From amphibian to mammal there is an increasing separation between the two sides of the heart.

The heartbeat is initiated when a wave of contraction spreads out from the "pacemaker," the *S-A node*, across the atria. This wave of contraction stimulates a second node, the *A-V node*, to send excitatory impulses down the fibers of the bundle of His, stimulating both ventricles to contract. When the ventricles contract (systole), the blood is forced out of the heart and into the arteries under high pressure. During the relaxation phase (diastole), the blood pressure in the arteries falls. This fluctuation (pulse) between the systolic and diastolic pressure diminishes in the arteries and no longer occurs in the capillaries and veins. In addition, friction causes the blood pressure to decrease as the blood moves farther away from the heart. Thus the blood pressure in the veins is not sufficient to move the blood back to the heart; the one-way valves, skeletal muscle action, and the motions of the chest during breathing aid in moving blood in the veins.

The enormous numbers, extensive branching, and small diameters of the individual capillaries ensure that all tissues are supplied with a large capillary surface area for the exchange of materials. Exchange may be accomplished by diffusion through the capillary cell walls, by transport in endocytotic vesicles, and by filtration between capillary cells.

The movement of materials into and out of the capillaries is governed by the balance between the hydrostatic blood pressure and osmotic pressure. At the arteriole end of a capillary the hydrostatic pressure usually exceeds the osmotic pressure, and water is forced out of the capillary; at the venule end the reverse is true, and water moves into the capillary. The balance between the osmotic and hydrostatic pressure is critical; changes in blood pressure or in the relative concentrations of proteins in the blood and tissue fluids can severely alter the balance of forces in the capillaries.

The *lymphatic system* helps maintain the osmotic balance of the body fluids by returning excess tissue fluid and proteins to the blood. These materials, absorbed into the permeable lymph capillaries, flow via lymph veins into the venous system. Lymph also transports the fats absorbed from the digestive tract. Lymph nodes located along the major veins act to filter out particles and also as sites of formation of some white blood cells.

Blood Blood consists of the *plasma*, the liquid portion of the blood, and the *formed elements*: red blood cells, white blood cells, and *platelets*. The water in the plasma is the solvent for the inorganic ions, plasma proteins, organic nutrients, nitrogenous waste products, hormones, and dissolved gases that are transported by the blood. The concentration of these solutes, especially the inorganic ions, remains relatively stable, being regulated by a variety of homeostatic mechanisms. *Homeostasis* (stability or equilibrium) is essential to the normal function of the organism.

Blood clotting is initiated when damaged tissue and disintegrating platelets release the substance *thromboplastin*, which converts the plasma protein *prothrombin* into *thrombin* (in the presence of Ca^{++} ions). The thrombin then converts the protein *fibrinogen* into *fibrin*; it is the fibers of fibrin that form the clot.

The red blood cells, or *erythrocytes*, contain the oxygen-carrying pigment *hemoglobin*. Each hemoglobin molecule can combine loosely with four molecules of O_2 to form oxyhemoglobin. When the partial pressure of O_2 is high (as in the lungs) the hemoglobin picks up oxygen; when the partial pressure of O_2 is low (as in the tissues) the oxyhemoglobin releases O_2. Both the pH of the blood and the temperature affect the affinity of hemoglobin for O_2. Among mammals, the oxyhemoglobin of smaller animals tends to release O_2 more readily than that of larger animals, which is consistent with the fact that smaller animals have a higher metabolic rate and that their tissues therefore require more O_2 per unit time. Because human fetal hemoglobin has a higher affinity for O_2 than maternal hemoglobin, it can obtain O_2 from the maternal hemoglobin.

The blood also transports carbon dioxide from the tissues to the lungs. Most of the CO_2 is carried in the form of the bicarbonate ion (HCO_3^-). The excess H^+ ions are bound to hemoglobin.

Leukocytes and their functions The five types of white blood cells, or *leukocytes*, defend the body against disease and infection. Some kinds of leukocytes act by carrying on phagocytosis, others by producing enzymes that detoxify dangerous substances, and still others by producing highly specific *antibodies* that destroy or inactivate certain kinds of foreign substances called *antigens*.

QUESTIONS

Testing recall

Determine whether each of the following statements is true or false. If it is false, correct the statement.

1 In a stem, the secondary phloem is located outside the layer of primary phloem.

2 As the stem of a woody plant grows through successive years, the cortex and pith become less and less important and are often virtually obliterated.

3 Phloem is the tissue of which wood is principally composed.

4 Mature functioning vessel cells are dead, but mature functioning tracheids are alive.

5 Mature sieve cells lack a nucleus but retain cytoplasm.

6 Mature xylem tracheid cells are metabolically more active than mature phloem sieve cells.

7 At present, most biologists believe transpiration pull is more important than root pressure in the upward movement of water in xylem.

8 The phloem of a tree can transport organic materials upward only.

9 Some mineral nutrients are much more mobile in the phloem than others.

10 Current evidence on the mechanism of transport in the phloem favors the mass-flow hypothesis over the cytoplasmic-streaming hypothesis.

11 Insects have blood capillaries but no red blood cells.

12 The walls of blood capillaries are composed of layers of muscle cells, connective tissue, and endothelium.

13 In a human being at rest, not all the blood is circulating at any given moment.

14 In fish, oxygenated blood goes directly from the gills to the peripheral circulation without first returning to the heart.

15 As an illustration of complete separation of the deoxygenated blood coming from the systemic circulation and the oxygenated

blood coming from the pulmonary circulation, a robin would be a better choice than a bullfrog.

16 A weak aorta is more likely to rupture during systole than during diastole of the cardiac cycle.

17 A reserve supply of blood is stored in the spleen in human beings.

18 At the arteriole end of a capillary, the osmotic pressure of the tissue fluid is greater than the blood pressure.

19 The lymphatic system returns excess tissue fluid to the blood.

20 Blood clotting begins as soon as the platelets are exposed to air.

21 In mammals, fetal hemoglobin has a stronger affinity for oxygen than adult hemoglobin.

22 Carbon monoxide is a dangerous poison because it has a strong tendency to bind to hemoglobin.

23 Iron is an essential component of the hemoglobin molecule.

24 Carbon dioxide is transported in the blood plasma principally as dissolved gas (CO_2).

25 Carbon dioxide accumulation tends to make the blood more acid.

Questions 26-34 refer to the diagram of the adult heart. Select the letter or letters on the diagram that correctly answer the questions.

26 A correct sequence for the flow of blood through the heart would be:

a A - B - C - E - D.   d B - A - D - E - C.
b A - B - D - E - C.   e C - E - D - B - A.
c D - E - C - B - A.

27 The heartbeat originates in the wall of what part of the heart?

28 Blood going to the lungs leaves from what part of the heart?

29 Blood from all over the body is returned to what part of the heart?

30 Which chamber pumps blood that will be distributed to all parts of the body?

31 The mitral or bicuspid valve is between which two chambers of the heart?

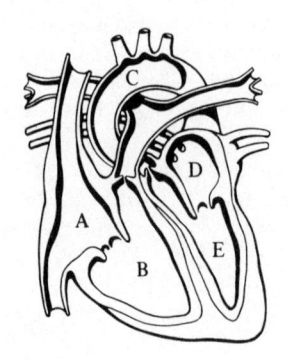

32 Which two chambers contract as a unit?

33 Which chambers pump oxygenated blood?

34 The pulmonary circulation connects which two chambers of the heart?

Study hint: A good way to study the human circulatory system is to trace the route a red blood cell follows as it moves through the circulatory system. Try the following:

35 List, in order, all the vessels and chambers of the heart that a red blood cell would pass through as it moves, by the most direct route possible, from

a capillaries in the lung to a vein in the brain

b an artery in the left leg to an artery in the left arm

Testing knowledge and understanding

Choose the one best answer.

1 Which of the following parts of the trunk of a large tree is the most important in transport of sucrose from the leaves to the roots?

a heartwood   d outer bark
b sapwood   e cambium
c inner bark

2 A nail is driven into the trunk of a young 6-foot-tall tree, 3 feet above the ground. Ten years later the tree is 30 feet tall. The nail will be how many feet above the ground?

a 3 feet   d 15 feet
b 6 feet   e 27 feet
c 10 feet

3 A beaver gnawed completely around a birch tree but did not proceed further to fell the tree. It was noticed that the leaves retained their normal appearance for several weeks, which were hot and sunny, but that the tree eventually died. It can be concluded that the most peripheral region it is *certain* that the beaver left functional was

a the phloem.
b the cortex.
c the xylem.
d the bark.
e the pith.

4 Which one of the following mature cells in a plant would be least active metabolically?

a root endodermal cell
b stem cortex cell
c mesophyll cell
d phloem companion cell
e xylem vessel element

5 Which one of the following statements is *not* true of a vessel cell in mature secondary xylem in a large tree?

a It was produced by cell division in the cambium.
b It once contained a nucleus, but no longer has one.
c Its cytoplasm is under the control of companion cells.
d Its thick cellulose walls are impregnated with lignin.
e Its end walls are extensively perforated.

6 Which of the following is *not* true of the xylem?

a Xylem tracheids and vessels fulfill their vital function only after their death.
b The cell walls of the tracheids are greatly strengthened with cellulose fibrils forming thickened rings or spirals.
c Water molecules are transpired from the cells of the leaf and replaced by water molecules in the xylem pulled up from the roots by the cohesion of water molecules.
d Movement of materials is by mass flow; materials move owing to a turgor-pressure gradient from "source to sink."
e In the morning sap in the xylem begins to move first in the twigs of the upper portion of the tree and later in the lower trunk.

7 Which of the following mechanisms could raise water to the greatest height in the xylem vessels of a tall tree?

a transpiration from the leaves combined with cohesion of water molecules
b root pressure
c cytoplasmic streaming
d active transport of water
e mass flow along a turgor-pressure gradient

8 Root pressure

a is not sufficient to raise water above ground level in any plant.
b is negative in all but the tallest trees.
c can push water to the top of a 10-foot corn plant but not to the top of a 200-foot tree.
d is the best explanation known for the transport of water to the tops of the tallest trees.
e is the driving force for the mass flow of sugar.

9 What is the principal pathway by which sugar travels in a plant?

a xylem vessels
b companion cells
c parenchyma cells
d sieve cells
e capillaries

10 According to the mass-flow hypothesis, materials are moved through the conducting cells as a result of

a a turgor-pressure gradient along the sieve tube.
b a pressure gradient along the sieve tube produced by cytoplasmic streaming.
c active transport between adjacent sieve cells.
d a flow along membranous interfaces within the sieve cells.
e a pull exerted by transpiration from the aerial parts of the plant.

11 Judging from what you know about animal circulatory systems, which one of the following animals would you expect to have the lowest blood pressure?

a rat
b frog
c robin
d grasshopper
e elephant

12 Which one of the following blood vessels carries deoxygenated blood in a mammal?

   a aorta
   b artery to the kidney
   c pulmonary artery
   d pulmonary vein
   e coronary artery

13 In which one of the following blood vessels of a horse would you expect to find the highest systolic blood pressure?

   a coronary artery
   b pulmonary vein
   c arteriole in the leg
   d capillary in the brain
   e anterior vena cava

14 In which one of the above vessels would there be the greatest difference between systolic and diastolic pressure?

15 A labeled red blood cell is released into the arterial circulation in the left leg. It is recaptured 30 seconds later in the left lung. What is the minimum number of chambers of the heart it must have passed through?

   a 0   d 3
   b 1   e 4
   c 2

16 Which one of the animals listed below would *not* show the pattern of circulation: heart ⟶ lung or gill capillaries ⟶ heart ⟶ systemic capillaries ⟶ heart?

   a dog   d turtle
   b frog   e bird
   c fish

17 In early 1973, Senator John Stennis was shot and severely wounded as he was getting out of his automobile on a Washington street. In a later account of the incident published in the *Washington Post*, there was the following passage:

The most serious wound, which at the outset many thought would cost him his life, was just at the belt-line on the left side. It affected his pancreas, colon, and portal vein, which supplies blood to the stomach. The vein was almost cut in two.

There is a biological error in this passage, which is that

   a the portal vein does not supply blood to the stomach.

   b the pancreas is not on the left side of the body.

   c the colon is nowhere near the indicated wound site.

   d a single bullet could not have hit both the pancreas and the colon.

   e the listed wounds would not have been sufficiently serious to endanger his life.

For questions 18-20 evaluate the statements using the following key:

   a The statement is true for *all* of the blood mentioned.

   b The statement is false for *all* of the blood mentioned.

   c The statement may be true of some of the blood mentioned but not of all of it.

   d There is no way to evaluate whether the statement is true or false.

18 Blood in a vein of a man's right arm must first go through the liver before it can go to the heart.

19 Blood in a capillary in the lung of a dog will go through the kidneys before going to the intestine.

20 Blood in the hepatic vein leaving the liver of a woman must first go through the kidney before it can go to the lungs.

21 A red blood cell leaves the left ventricle of a human and travels into an artery leading to the right arm. How many capillary beds will it pass through before it returns to the left ventricle?

   a none   d three
   b one   e four
   c two

22 The diagram of a capillary shows the hydrostatic blood pressure measured at three points along this particular capillary. Net outward movement of water at the arteriole end and net inward movement at the venule end would be facilitated if the osmotic pressure of the blood (relative to the tissue fluids) were

   a 15 mm Hg.
   b 20 mm Hg.
   c 30 mm Hg.
   d 40 mm Hg.
   e 50 mm Hg.

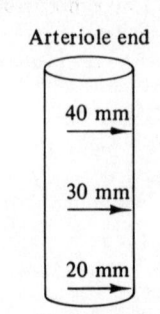

Arteriole end
40 mm
30 mm
20 mm
Venule end

23 Assume that the graph describes the pulse pressure and cross-sectional area of a circulatory system whose dynamic interrelationships are similar to those of the corresponding human system. Considering only these two parameters, what would you expect a graph of velocity to look like in such an organism?

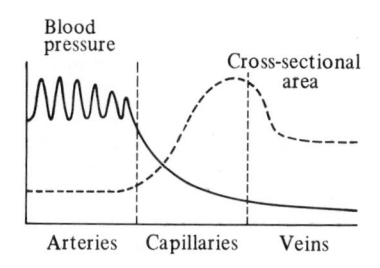

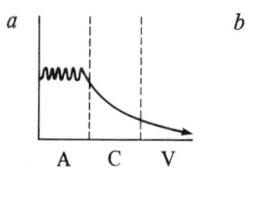

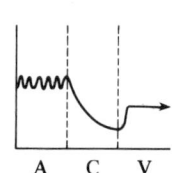

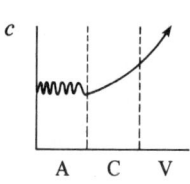

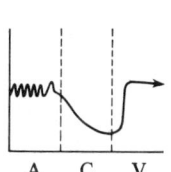

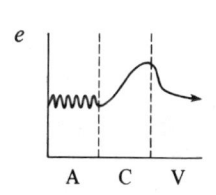

24 All *but one* of the following are parts of or contained in the lymphatic system. That part is the

a capillary networks.
b arterioles.
c valves.
d leukocytes.
e nodes.

25 Which one of the following would you *least* expect to find in lymph?

a salt
b lipid
c water
d protein
e hemoglobin

26 Which one of the following statements about blood clotting is *not* true?

a For clotting to occur, the blood must be exposed to molecular oxygen.
b When platelets are mechanically damaged, they release a substance that initiates clotting.
c For clotting to occur, calcium ions must be present.
d The plasma protein fibrinogen is necessary for blood clotting.
e The plasma protein prothrombin is necessary for blood clotting.

27 The graph shows dissociation curves for the hemoglobin of five different animal species. On the basis of the characteristics of its hemoglobin, which animal would be best adapted for life at high altitudes?

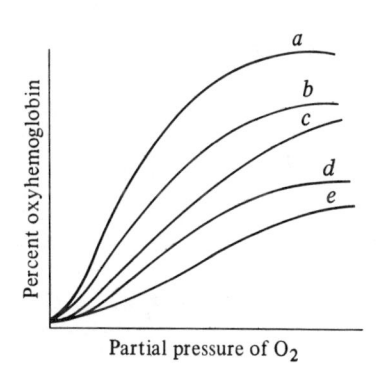

28 The typical S-shaped dissociation curve for hemoglobin owes its shape to the fact that

a oxyhemoglobin releases O_2 when the partial pressure of O_2 is high and picks up O_2 when the partial pressure is low.
b an increase in the partial pressure of O_2 favors the release of O_2.
c the binding of the first molecule of O_2 facilitates the binding of subsequent molecules.
d oxyhemoglobin is formed only when the partial pressure of O_2 is very high.

29 C^{14}-labeled CO_2 released by a mammalian cell would probably be found moving in the blood primarily as

a carboxyhemoglobin.
b bicarbonate ion.
c carbonic acid.
d acid hemoglobin.
e oxyhemoglobin.

For further thought

1. As plants evolved to land forms, they were confronted with the problem of support, since they no longer had water to support their weight. How was this problem solved?

2. Trace the route a molecule of water would follow from the place where it enters a vascular plant until it is lost from the leaves by transpiration.

3. Explain how the carbohydrates synthesized in the leaves get from the leaves to the root and other parts of the growing plant.

4. Distinguish between open and closed circulatory systems. What are the advantages of a closed system?

5. Why is the type of heart seen in the fish and primitive vertebrates inadequate for most higher vertebrates? How has the circulatory plan been modified in the higher vertebrates?

6. If a person touched a high-voltage electric power line, what would be the effect on his heart? Explain.

7. A man goes to his doctor and finds that his blood pressure is 190/110. What do these numbers refer to? What is the effect of high blood pressure on the exchange of materials in the capillaries? What symptoms might result over a long period of time if this condition remains untreated?

8. What roles are played by hemoglobin in addition to the role it plays in oxygen transport?

9. John Jones was admitted to the hospital with a swollen and painful infected finger. There was a red streak up his arm, and several large, tender lumps in his armpit.

    a What caused the red streak and the lumps under his arm?

    b State the factors that contributed to the swelling.

    c Discuss the ways in which the human body protects itself against invading bacteria and viruses.

Chapter 8

REGULATION OF BODY FLUIDS

KEY CONCEPTS

1 The cells of multicellular animals require a relatively nonfluctuating fluid environment in order to carry out their life functions.

2 Living organisms have various mechanisms for maintaining a relatively constant internal environment despite changes in the external environment. This constant internal environment is a prerequisite to their ability to live in a variety of environments.

3 The extracellular fluid of plants differs greatly from that of animals in that the extracellular fluid of plants is not fully distinct from environmental water and cannot be as well regulated as the tissue fluid and blood of animals.

4 Plant cells can withstand much greater fluctuations in the makeup of the fluids bathing them than animal cells.

5 Multicellular animals have various mechanisms for ridding their bodies of metabolic wastes and for regulating their salt and water balance to maintain the constancy of their body fluids.

OBJECTIVES

After studying this chapter and reflecting on it, you should be able to

1 State the **principle of homeostasis and** explain why the body's health and survival depend upon the maintenance of homeostasis.

2 Contrast the extracellular fluids of multicellular plants with those of animals.

3 Explain why plant cells seem able to withstand greater fluctuations in the makeup of their extracellular fluids than animal cells.

4 Discuss what happens to typical plant and animal cells in a hypotonic environment, then explain what happens in a hypertonic environment.

5 Explain the functional **significance of the** hepatic portal system.

6 Discuss the role of your liver in maintaining homeostasis. Include in your answer the liver's function in carbohydrate, protein, and fat metabolism.

7 Define excretion and osmoregulation and state the difference between excretion and elimination.

8 List the three main nitrogenous waste products, indicate where each is produced, and give the advantages and disadvantages of each substance as a waste product.

9 Give the main nitrogenous waste product for each of the following animals: jellyfish, planaria, Protozoa, marine fish, freshwater fish, frog, snake, robin, human being.

10 Contrast the osmoregulatory problems faced by freshwater bony fish with those of marine bony fish and with those of sharks. Discuss the adaptations that enable these animals to cope with these problems.

11 Compare and contrast the excretory mechanisms of protozoans, flatworms, annelid worms, insects, and mammals, relating these mechanisms to the type of circulatory system each possesses.

12 Discuss in some detail the process of urine formation in man, explaining

   a the source of urea, and when the urea content of the blood will be highest;
   b the roles of filtration, reabsorption, and tubular excretion;
   c specifically where and how each of these processes occurs.

13 Explain how the countercurrent-multiplier system of the loop of Henle acts to concentrate the urine.

14 Discuss the role played by your kidney in maintaining homeostasis.

15 Describe a model for the sodium-potassium pump and explain how such a pump might work in the cells of the gills of a freshwater fish.

SUMMARY

Living cells require a relatively constant environment to carry out their life functions. Life is believed to have begun in the ancient seas. Here the primitive cells were bathed in seawater of relatively constant composition. As more complex multicellular animals arose, body fluids developed that could provide the same type of stable environment for the internal cells. A variety of mechanisms evolved for maintaining the *homeostasis* of the body fluids despite changes in the external environment.

The extracellular fluids of plants are not separate from the environmental water. Consequently plants regulate only the composition of their intracellular fluids whereas animals must regulate the composition of both intracellular and extracellular fluids. Plant cells can withstand much greater fluctuations in their extracellular fluids than animals. Because the rigid cell wall resists expansion, the plant cell can withstand pronounced changes in the osmotic concentration of the surrounding fluids as long as the fluids remain hypotonic to the cell. If the external fluids become decidedly hypertonic to the cell, the cell may lose water and shrink away from the wall (*plasmolysis*). The presence of the cell wall in plants and its absence in animals thus makes the problem of salt and water balance quite different in plant and animal cells.

Land plants are frequently exposed to conditions that may cause excessive water loss by evaporation. Plants have evolved many different types of adaptation to resist desiccation.

The vertebrate liver The liver is important in maintaining a relatively constant environment for the body cells. All the blood from the intestine and stomach is collected in the *portal vein* and conducted to the liver, where it flows into a second set of capillaries before it is collected into the *hepatic veins* and emptied into the posterior vena cava. Because of this portal system, the products of digestion are brought directly to the liver cells, where the levels of these digestive products can be regulated.

When the incoming blood is high in glucose, the liver removes the excess and stores it as glycogen or fat; when the incoming blood is low in glucose, the liver reconverts glycogen into glucose to maintain the blood-sugar level. The liver is the center for fat metabolism; fatty acids and other lipid materials are processed here and some plasma lipids are synthesized.

The liver also removes many of the amino acids from the blood, temporarily storing small quantities and gradually returning them to the blood. Excess amino acids are *deaminated* (the amino group is removed) and the remainder of the molecule is converted into carbohydrate or fat. In the deamination process the amino group ($-NH_2$) is converted into *ammonia* (NH_3). The ammonia may be released directly as a waste product or it may be converted into the less toxic compounds *urea* and *uric acid*. These are released into the blood and must be removed from the body.

In addition to its metabolic functions, the liver also detoxifies many injurious chemicals, manufactures many plasma proteins, stores certain substances, destroys red blood cells, synthesizes bile salts, and excretes bile pigments.

The problem of excretion and salt and water balance in animals Animals need mechanisms to rid their bodies of metabolic wastes (*excretion*) and to regulate their salt and water balance.

The nitrogenous excretory product characteristic of most aquatic animals is ammonia. Ammonia is poisonous but is easily released from the body if the water supply is plentiful. Many marine invertebrates simply release their wastes across their membrane surfaces. Such organisms are isotonic with their environment and thus have no problems with water balance as long as they remain in the sea. Marine organisms exposed to a hypotonic environment tend to lose salts and gain water; some have evolved adaptations to regulate their osmotic concentration and thus can survive in a hypotonic environment.

Freshwater animals have an osmotic concentration that is less than that of seawater but greater than that of freshwater. A freshwater organism tends to gain water and lose salts to the surrounding water. To survive, the organism must have an osmoregulatory structure to pump out the excess water and some mechanism to absorb salts. Freshwater bony fishes compensate by drinking very little, by actively absorbing salts through specialized gill cells, and by excreting copious dilute urine.

Marine bony fishes evolved from freshwater fishes and thus are hypotonic relative to seawater. They tend to lose water and take in salt. They compensate by drinking constantly, by actively excreting salts across their gills, and by producing little urine.

The marine elasmobranchs (sharks, etc.) compensate in a different way—by maintaining a high concentration of urea in their blood so that their total osmotic concentration is slightly greater than that of seawater.

The greatest problem for a land animal is desiccation. The water lost by evaporation, elimination, and excretion must be replaced. Amphibians and mammals produce urea as their nitrogenous waste product; it must be released from the body in solution, thus draining away needed water. Most reptiles, birds, insects, and land snails excrete uric acid. Little water is lost in the excretion of this highly insoluble compound. The uric acid produced by a developing embryo is precipitated and stored in the eggs of these animals; in this form it is not harmful. Uric acid excretion is correlated with egg laying, whereas urea excretion is correlated with giving birth to living young.

Excretory mechanisms in animals Many unicellular and simple multicellular animals have no special excretory structures; the nitrogenous wastes are excreted across membrane surfaces. Some Protozoa do have an excretory organelle, the *contractile vacuole*, which collects fluid and expels it from the cell. Its primary function is the elimination of excess water.

Flatworms have a simple tubular type of excretory system, the *flame-cell system*, to eliminate excess water. Water and some waste materials move into the bulblike flame cells, whose cilia move the liquid through a series of tubules to several excretory pores.

In animals that have evolved closed circulatory systems, the blood vessels have become intimately associated with the excretory organs, making possible direct exchange of materials between the blood and the excretory system. In earthworms, each segment of the body has a pair of excretory organs called *nephridia*, which open to the outside. Tissue fluids move into the open end of this tubular structure and pass through the coiled tubule to the exterior. Blood vessels associated with the tubules also secrete substances into the tubules for excretion.

The vertebrate kidney The vertebrate excretory organ is the kidney, which is composed of many functional units called *nephrons*. Three processes are important in urine formation: ultrafiltration, reabsorption, and tubular excretion. The *Bowman's capsule* of each nephron receives fluid from a tuft of capillaries called the *glomerulus*. High blood pressure within the glomerulus forces small molecules from the blood into the Bowman's capsule (ultrafiltration). As the filtrate moves through the tubules of the nephron most of the water and substances needed by the body are reabsorbed into a second set of capillaries associated with the tubules. In addition, some chemicals are actively removed from the blood by the tubules and deposited in the filtrate (tubular excretion). The nephrons empty into *collecting tubules*; from here the urine is conducted into the kidney *pelvis*, then down the *ureter* to the urinary bladder for storage, and finally out the *urethra*.

The portion of the nephron called the *loop of Henle* makes it possible to produce concentrated urine; it establishes a sodium gradient in the tissue fluid of the kidney medulla that favors the osmosis of water out of the collecting tubules into the surrounding tissue fluid.

Most substances reabsorbed by the kidney have a *threshold level*; if the concentration of such a substance exceeds the threshold, the excess is not reabsorbed by the tubules but instead appears in the urine. Thus, the kidneys play a vital role in maintaining homeostasis; they help to regulate the blood-sugar level and the concentration of various inorganic ions in the blood.

The kidneys of vertebrates vary in their ability to produce concentrated urine. Marine fishes, turtles, and birds cannot produce concentrated urine; they excrete excess salt by special cells or glands. Seals and fish-eating whales have kidneys capable of excreting a urine with high urea concentration. Kangaroo rats have extraordinarily efficient kidneys capable of producing extremely concentrated urine. The human kidney is incapable of producing urine with a very high concentration of either salt or urea.

The excretory organs of insects are called *Malpighian tubules*. They are diverticula of the digestive tract located at the junction between the midgut and hindgut. The tubules are bathed directly by the blood; fluid from the blood moves into the tubules, where the uric acid is precipitated. The concentrated urine moves into the hindgut and rectum, where it is exposed to a powerful water-absorptive action; the urine and feces leave the rectum as dry material.

The processes of excretion and osmoregulation depend on active transport. Much energy is required to accumulate some substances and expel others. A model for the so-called sodium-potassium pump has been suggested. The model proposes a protein complex in the cell membrane (called Na^+K^+–$ATPase$) that acts as both the permease for the transport of Na^+ and K^+ and the enzyme for the hydrolysis of ATP. The process depends on conformational changes in this protein. In one of its conformations (E_1) the protein has an affinity for Na^+ and ATP inside the cell; binding of these induces the molecule to change shape, moving Na^+ to the exterior, where it is released. In this conformation (E_2) the protein has an affinity for K^+; when the K^+ binds, the protein returns to its original (E_1) conformation, and K^+ is released inside the cell. The energy for the conformational changes is provided by the hydrolysis of ATP.

Many epithelial cells are involved in active transport in osmoregulatory and excretory organs; these cells probably utilize the same basic type of pump.

QUESTIONS

Testing recall

Complete the following statements with the words greater than, less than, *or* equal to.

1 If you were to measure the extracellular fluid of a marine alga living in seawater you would find its osmotic concentration is _____ seawater. The osmotic concentration of intracellular fluid of the alga would be _____ that of the extracellular fluid.

2 The osmotic concentration of the extracellular fluid of a marine jellyfish living in the ocean is _____ that of seawater. The osmotic concentration of its intracellular fluid is _____ that of its extracellular fluid.

3 If you were to measure the osmotic concentration of the extracellular fluids of a marine bony fish you would find its osmotic concentration to be _____ seawater, and the osmotic concentration of the interior of its red blood cell would be _____ that of the blood.

4 The osmotic concentration of the extracellular fluid in the cortex of a root of a land plant would be _____ that of soil water and the osmotic concentration of the cortex cells would be _____ soil water.

Listed below are the approximate osmotic concentrations of seawater and freshwater, and of the body fluids of various animal groups living in these environments.

	percent
Seawater	3.5
a marine sharks	3.6
b marine invertebrates	3.5
c marine bony fish	1.5
Freshwater	0.01–0.3
d freshwater invertebrates	0.4–0.6
e freshwater bony fish	0.85
f amphibians	0.85
g mammals	0.9
h reptiles	0.9

Using this information, answer the following questions:

5 Which organisms will tend to take in excessive water from their environment?

6 Which organisms are essentially isotonic with their environment?

7 Which organisms will tend to lose too much body water to their environment?

8 Which organisms tend to take in too much salt from the environment?

9 Which organisms have special mechanisms for actively taking up salts from the environment and releasing them into their body fluids?

10 Which organisms probably use ammonia as their main nitrogenous waste product?

11 Which organisms use mainly urea as their nitrogenous waste product?

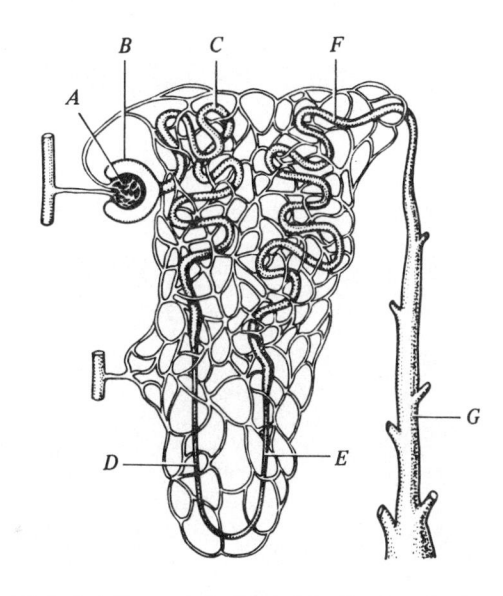

12 Label the parts indicated in the sketch of the nephron.

13 Referring to the sketch, indicate by letter in which part of the nephron each process occurs.

   a ultrafiltration of the blood
   b passive transport of Na$^+$
   c reabsorption of most water
   d pumping of Na$^+$ out of the tubule

14 Which parts of the nephron are found in the cortex of the kidney?

Testing knowledge and understanding

Choose the one best answer.

1 A plant cell placed in a hypotonic environment will

   a lose water and become flaccid.
   b actively transport salts out of the cell.
   c take up water and burst.
   d take up water and become turgid.
   e become impermeable to prevent water loss.

2 A biologist experimenting with a protozoan noticed that the contractile vacuole stopped contracting although the other parts of the organism seemed healthy and active. Which

of the following experiments was the one most likely to have produced this result?

- *a* transferring the protozoan from a lighted environment to a dark environment
- *b* transferring the protozoan from freshwater to seawater
- *c* transferring the protozoan from seawater to freshwater
- *d* changing the pH of the medium from 7.0 to 6.5
- *e* cooling the medium from 20°C to 15°C

3 A trout in a freshwater stream

- *a* loses water and salts by diffusion, and compensates by drinking large quantities of water and actively absorbing salts via the gills.
- *b* takes in much water and loses salts by diffusion, and compensates by excreting copious dilute urine and actively absorbing salts via the gills.
- *c* loses water and takes in large quantities of salts by diffusion, and compensates by actively absorbing water via the gills and by excreting salty urine.
- *d* takes in much water and salt by diffusion, and compensates by excreting copious urine and actively excreting salts via the gills.
- *e* takes in much water and loses salt by diffusion, and compensates by nearly constant drinking and by actively excreting both water and salts via the gills.

4 A freshwater fish is placed in a marine environment. What would you expect to happen to this fish eventually?

- *a* It will dehydrate.
- *b* It will take up water.
- *c* It will neither gain nor lose water.
- *d* It will be in osmotic balance with the seawater.
- *e* It will actively excrete salt.

5 Dehydration is *not* a problem for which one of the following animals?

- *a* saltwater fish
- *b* freshwater fish
- *c* camel
- *d* human
- *e* desert iguana

6 An erythrocyte is in an artery leading to the duodenum of a human being. How many capillary beds will it probably pass through before it reaches the right ventricle of the heart?

- *a* one
- *b* two
- *c* three
- *d* four
- *e* five

7 Shortly after a person has eaten a large meal, in which of the following blood vessels would you expect the blood to contain the highest concentration of sugar?

- *a* aorta
- *b* posterior vena cava
- *c* renal artery
- *d* hepatic portal vein
- *e* jugular vein

8 In the human being, blood leaving the liver in the hepatic vein ordinarily has a higher concentration of _____ than other blood.

- *a* bile
- *b* urea
- *c* oxygen
- *d* erythrocytes
- *e* leukocytes

9 Which of the following is *not* a function of the human liver?

- *a* to remove the excess glucose from the blood and store it as glycogen
- *b* to inactivate certain drugs and injurious chemicals
- *c* to deaminate amino acids and convert the ammonia thus formed into urea
- *d* to manufacture many plasma proteins and some plasma lipids such as cholesterol
- *e* to secrete digestive enzymes into the small intestine

10 You are given an unknown animal to study. It is bilaterally symmetrical and has a complete digestive system. It has a circulatory system but the blood carries very little oxygen. The animal's principal nitrogenous excretory waste product is uric acid. Which one of the following animals best fits this description?

- *a* salmon
- *b* earthworm
- *c* sparrow
- *d* planarian
- *e* grasshopper

11 The nitrogen present in proteins can be eliminated in the form of ammonia, which is highly toxic and must be removed from the body rapidly. In which of the following habitats would you be most likely to encounter organisms using ammonia as their nitrogenous waste product?

- *a* underground
- *b* in water
- *c* on land
- *d* in the air
- *e* in deserts

12 Hydrostatic pressure powers the process of

- *a* filtration across the glomerulus.
- *b* reabsorption of water and dissolved substances at the venule end of the capillary.

c sodium gradient maintenance in the kidney tubule.
d salt and glucose reabsorption in the kidney tubules.
e Na⁺ recycling in the loop of Henle.

13 Based on your knowledge of the anatomy and physiology of the kidney, in which of the following blood vessels would you expect to find the highest concentration of plasma proteins?

a the renal artery
b the arteriole entering Bowman's capsule
c the vessel leaving Bowman's capsule
d the venule leaving a nephron
e the renal vein

14 The liquid collected by Bowman's capsules in a human kidney may best be characterized as

a concentrated urine.
b dilute blood.
c blood plasma minus blood proteins.
d a solution of urea.
e blood minus formed elements.

15 Assume that you have dissected kidneys from the following mammals and measured the lengths of the loops of Henle and the collecting ducts in each. In which animal would you expect to find these structures to be proportionately longest in comparison to the overall size of the animal?

a a desert rodent   d an elephant
b a marine bird   e a forest rabbit
c a marine seal

16 In order for humans to drink saltwater, the hairpin loops of the kidney tubules would have to be

a longer.
b shorter.

17 The *second* of the two capillary beds in the kidney

a reabsorbs nutrients.
b is called the glomerulus.
c ultrafilters the blood.
d goes to the liver.
e carries less concentrated blood than does the first bed.

18 A person with kidney failure who is undergoing blood dialysis would probably be restricted to a diet low in

a protein.   d calories.
b carbohydrate. e bulk.
c fat.

19 The kidneys of vertebrates function in all of the following ways but one. This exception is

a excretion of metabolic wastes.
b regulation of water concentration.
c regulation of ion concentration.
d elimination of undigested wastes.
e elimination of materials in the blood that are in oversupply.

20 Large molecules such as uric acid, creatinine, and penicillin require ATP energy to be excreted by the kidney. These substances enter the nephron by the process of

a ultrafiltration.   d diffusion.
b osmosis.   e tubular secretion.
c reabsorption.

21 According to the countercurrent-multiplier model for water removal from the kidney tubules

a Na⁺ is being pumped out of the ascending loop of Henle.
b Na⁺ is being pumped out of the descending loop of Henle.
c water is actively transported out of the collecting tubules.
d the filtrate becomes hypertonic to the tissue fluid.
e water is actively reabsorbed in the convoluted tubules.

22 According to the current model of the sodium-potassium pump,

a two Na⁺ ions are pumped into the cell for every three K⁺ ions that are pumped out.
b hydrolysis of ATP is required for the step in which K⁺ ions are pumped into the cell.
c phosphorylation of the permease induces a conformational change that opens a channel through the membrane.
d separate permeases are required for the transport of Na⁺ and K⁺ ions.
e two ATP molecules are required for each cycle of the pump.

23 Which of the following processes that occur in the kidney directly requires active transport?

a movement of Na⁺ into the descending limb of the loop of Henle
b movement of Na⁺ out of the ascending limb of the loop of Henle
c filtration into Bowman's capsule
d osmosis of water in the collecting tubule
e movement of urea into the nephron

For further thought

1 Lake Baikal in the Soviet Union is the only known habitat of freshwater seals; all other seal species inhabit marine environments. What changes are most likely to have occurred in the kidney functions of these freshwater seals?

2 A patient was admitted to a hospital with a preliminary diagnosis of viral hepatitis (inflammation of the liver). The following tests were performed:

   a Fasting blood sugar. The blood sugar level was measured six hours after an injection of glucose. Result: blood sugar was 58 mg/100 ml. (Normal: 70–120 mg/100 ml.)
   b Serum albumin test. The level of albumin in the blood was determined. Result: serum albumin was lower than normal.
   c Prothrombin time. The time it takes for blood to clot was measured. Result: clotting time was higher than normal.
   d Bromsulfalein test. This dye was injected into the blood and the amount of dye retained was measured. Result: 50 percent of the dye was retained in 30 minutes. (Normal: 10 percent or less.) (This dye is not excreted by the kidneys.)

   On the basis of what you know about liver function, explain the above results. What symptoms do you think the patient might present?

3 A molecule of water is absorbed into a capillary in the small intestine of a human being. Trace the route this molecule will follow as it goes by the shortest possible route to the urinary opening. (Name all the blood vessels and parts of the kidney, etc.)

4 Would the urine composition of a vegetarian differ from that of a heavy meat eater?

5 Approximately 180 liters of fluid are filtered through the kidney every 24 hours, but only about 1.5 liters of urine are produced. What is the average rate of filtration in ml/minute? (1 liter = 1000 ml.) What will happen to the rate of filtration if the blood pressure is increased? What is the rate of urine production in ml/minute? If the two kidneys together have two million glomeruli, how many ml of fluid are filtered by each glomerulus in 24 hours?

   a 9.0 ml   d 8.9 ml
   b 0.9 ml   e 0.89 ml
   c 0.09 ml   f 0.089 ml

6 What do kidneys and nephridia have in common? Why must insects have an excretory system that operates on an entirely different principle?

7 Describe a model that would account for the movement of Na^+ out of the ascending loop of Henle.

Chapter 9

CHEMICAL CONTROL

KEY CONCEPTS

1. Chemical control mechanisms play an important role in the coordination of the myriad functions of living organisms. Hormonal control is common to plants and animals.

2. Plant hormones are primarily involved in growth and development whereas animal hormones mediate a great variety of functions in addition to growth and development.

3. In multicellular animals, the nervous and endocrine control systems are intimately related; the nervous system can influence the activity of the endocrine system. Nervous and hormonal control can be regarded as parts of a single, integrated control system.

4. The secretion of many animal hormones is regulated by a negative feedback control mechanism in which low concentrations of a particular hormone stimulate increased secretion of that hormone. When the concentration reaches a certain level in the blood, hormonal secretion is reduced.

5. Many hormones act on target cells indirectly, by interacting with specific receptors on the cell membrane, thus activating cytoplasmic **enzyme systems** in the cell. Other hormones act more directly, by moving into the target cell and interacting with the genetic material or with the process of protein synthesis.

OBJECTIVES

After studying this chapter and reflecting on it, you should be able to

1. Define hormones and indicate how they are transported from the site of synthesis to the sites of action in higher animals.

2 Give at least three ways in which plant hormones differ from animal hormones.

3 Discuss the roles of auxin in the control of plant growth and explain some of the experiments used to gain knowledge on this subject.

4 Explain why plant stems are positively phototropic and negatively geotropic whereas plant roots are negatively phototropic and positively geotropic.

5 List a major function of each of the following hormones or groups of hormones: gibberellins, cytokinins, ethylene, abscisic acid, florigen.

6 Explain how auxin is involved in apical dominance.

7 Contrast the effect of auxin and gibberellins on cell elongation in the plant stem.

8 Explain how auxins and cytokinins influence cell division and growth.

9 Discuss the roles of hormones in the development and ripening of fruit.

10 Discuss the role of the plant growth inhibitors in maintaining dormancy and preparing the plant for overwintering. Indicate how dormancy is broken.

11 Describe the process of leaf abscission and how auxin, abscisic acid, and ethylene influence this process.

12 Differentiate among short-day plants, long-day plants, and day-neutral plants.

13 Define photoperiodism and discuss the relationship among florigen, phytochromes, and photoperiodism in the control of flowering.

14 Explain how juvenile hormone, brain hormone, and molting hormone interact to control the molting process in some insects.

15 Define the term endocrine and list the major endocrine organs in the human body.

16 Explain how hormones act to coordinate the secretion of enzymes in the digestive process.

17 Give the standard procedure for determining whether an organ has an endocrine function.

18 List four actions of insulin in reducing the concentration of glucose in the blood. Contrast its action with that of a second hormone from the pancreas, glucagon.

19 Discuss functions of the hormones of the adrenal medulla, adrenal cortex, thyroid, and parathyroid.

20 Describe the relationship between the hypothalamus and the posterior pituitary.

21 Give the functions of prolactin and growth hormone.

22 Describe the roles of the anterior pituitary and hypothalamus in coordinating the activity of the vertebrate endocrine system.

23 Contrast the function of the pineal in lower vertebrates with its function in higher vertebrates.

24 Describe the two-messenger model of hormonal control and show how this model explains why only small quantities of hormone are required for normal functioning and why only certain cells are influenced by a given hormone.

25 Discuss the mode of action of the steroid hormones.

26 Describe the group of substances called prostaglandins; indicate where they are produced and how they might act.

27 Distinguish between self-fertilization and cross-fertilization and indicate which is most common among animals.

28 Explain how external fertilization differs from internal fertilization and relate the differences to the number of gametes produced, the necessity for water, and the type of environment inhabited.

29 Summarize the characteristic reproductive methods of the major classes of vertebrates.

30 Draw a diagram of the embryonic membranes in a bird or reptilian egg.

31 List in order all the structures through which a sperm passes between its point of origin in the human male and the exterior. Indicate where the accessory sexual glands add their secretions.

32 Give the components of the semen and list four functions of the seminal fluid.

33 Name the structures of the human female genital system and give a function for each. Indicate where fertilization occurs.

34 Describe the sequence of changes in one human menstrual cycle, including the hormonal changes and interactions involved.

35 List five methods of birth control and explain how each method works.

36 Indicate the role played by each of the following in controlling or maintaining pregnancy in the human female: placenta, corpus luteum, chorionic gonadotrophin, progesterone.

37 List the hormones thought to be involved in the birth process and in milk production; give the function of each hormone.

SUMMARY

Plant hormones are produced most abundantly in the actively growing parts of the plant body: the apical meristems, young growing leaves, and developing seeds. The known plant hormones are primarily involved in regulating growth and development.

Auxins are a class of plant hormones that produce a variety of effects, the most important of which is control of cell elongation. Experiments show that auxin is produced by the apical meristem and moves downward, promoting cell elongation in the stem.

Light influences the distribution of auxin within the stem by causing migration of the auxin away from the lighted side. Consequently the illuminated side of the plant grows more slowly than the shaded side and the plant turns towards the light—a phenomenon called *phototropism*. Gravity also causes an unequal distribution of auxin within the stem and root. The plant shoot turns away from the pull of gravity, showing *negative geotropism*, whereas the root exhibits *positive geotropism*. This is believed to result from a difference in auxin sensitivity between the two organs.

Auxin produced in the terminal bud moves downward in the shoot and inhibits the development of the lateral buds (*apical dominance*). Auxin produced by the pollen grain and developing seed stimulates the development of the floral ovary into the fruit. Auxin also acts to prevent the abscission (dropping) of flowers, fruits, and leaves by inhibiting formation of the *abscission layer*. In early spring it stimulates renewed cell division in the cambium. It also initiates the formation of lateral roots from the root pericycle and promotes the development of adventitious roots from cuttings. Auxin-like chemicals such as 2,4-D and 2,4,5-T are used as broad-leaved weed killers.

The most dramatic effect of another class of plant hormones, the *gibberellins*, is stimulation of rapid stem elongation in dwarf plants. Because gibberellin can move freely throughout the plant, most of its effects on the pattern of growth are different from those of auxin. Gibberellins are also active in breaking dormancy, inducing the formation of a starch-hydrolyzing enzyme in germinating seeds, stimulating flowering in some biennials, inducing flowering in some long-day plants, and stimulating fruit set.

Cytokinins are a class of hormones that promote cell division. In the normal growing plant, cytokinins and auxins may act synergistically in some situations and antagonistically in others; e.g. synergistically in promoting DNA replication and antagonistically on the growth of lateral buds. Cytokinins also play a role in chloroplast development, breaking dormancy in some seeds, enhancing flowering, promoting fruit development, and delaying the aging of leaves.

Auxins, gibberellins, and cytokinins are growth-promoting hormones, but plants also produce some growth inhibitors, the most important of which is the hormone *abscisic acid*. Inhibitors are important in maintaining dormancy of buds and seeds when conditions for growth are unfavorable, as in winter or very dry periods. Dormancy is broken when the inhibitors have been degraded, destroyed, or leached away by water. Abscisic acid also induces changes that prepare the plant for overwintering.

The hormone *ethylene* is a gas that induces the onset of the climacteric in fruit, causing ripening. It also contributes to leaf abscission and various other changes involved in the aging process.

Control of flowering The flowering response in many plants is initiated by *photoperiodism*, the length of light and dark periods. Most plants can be classified as: (1) short-day plants, which flower only when the night is *longer* than a critical value; (2) long-day plants, which flower only when the night is *shorter* than a critical value; and (3) day-neutral plants, which are independent of day and night length. The critical night length is a *maximum* value for a long-day plant and a *minimum* value for flowering in a short-day plant.

It is believed that an inducing photoperiod causes the leaves to produce the hormone *florigen*, which moves through the phloem to the buds and stimulates flowering. A noninducing photoperiod causes the leaves of many plants to inhibit florigen in some way. All attempts to isolate florigen have failed.

Light is detected by a receptor pigment in the plasma membrane called *phytochrome*, which exists in two forms, one that absorbs red light (P_r) and one that absorbs far-red light (P_{fr}). The red and far-red light interconverts the phytochrome between its two forms. P_{fr} is the less stable form in the dark; it reverts to P_r over time, and some of it is enzymatically destroyed.

When phytochrome is exposed to both red and far-red light simultaneously (as in sunlight) the red light dominates and most of the pigment is converted into P_{fr}. During the night the P_{fr} supply dwindles as a result of reversion and destruction. Thus the plant has a way to detect whether it is day or night.

The mechanism that enables the plant to measure the length of the dark period is apparently tied to an "internal clock." The phytochrome mechanism determines whether it is day or night while the internal clock measures the length of the dark period. Once these have indicated to the plant that the photoperiod is appropriate, florigen is produced and flowering is initiated.

Phytochrome participates in many other light-induced functions such as the germination of seeds, gibberellin-controlled stem elongation, expansion of new leaves, breaking of dormancy, and the formation of plastids.

Hormones in animals The tissues that produce and release hormones in animals are termed *endocrine* tissues. The hormones are secreted directly into the blood, which then transports them to other parts of the body.

Hormonal control mechanisms have been found in a variety of invertebrates; those in insects have been most extensively studied. In invertebrates the nervous system and endocrine function are usually intimately associated.

In mammals, hormones produced by the mucosal lining of the stomach and small intestine act to coordinate the digestive process. The Russian physiologist Pavlov showed that gastric secretion is partly controlled by hormones; partially digested food in the stomach stimulates the gastric mucosa to release the hormone *gastrin* into the blood, where it is carried to the gastric glands and stimulates the secretion of gastric juice. Likewise, the presence of food in the small intestine stimulates the intestinal lining to release the hormones *secretin* and *cholecystokinin*. Secretin stimulates the secretion of pancreatic juice by the pancreas; cholecystokinin the release of bile from the gall bladder. The *islet cells* of the pancreas secrete *insulin*, which acts to reduce the blood-glucose concentration by stimulating glucose absorption in muscle and adipose cells, by promoting glucose oxidation and glycogen synthesis in liver and muscle cells, and by inhibiting glycogen hydrolysis. It also promotes the synthesis of fat and proteins while inhibiting their breakdown.

Insulin deficiency results in the disease diabetes mellitus. Glucose is not removed from the blood and the blood-sugar level rises above normal; some of the excess glucose appears in the urine. Yet the body lacks sufficient energy because of the impairment of carbohydrate metabolism. Consequently, cells begin to metabolize fats and proteins, producing some toxic substances, which affect the delicately balanced pH of the body.

The pancreas secretes another hormone, *glucagon*, which has effects opposite to those of insulin; it causes an increase in the blood-glucose concentration.

The two *adrenal glands*, located above the kidneys, consist of an inner *medulla* and outer *cortex*, which remain functionally distinct.

The adrenal medulla secretes two hormones, *adrenalin* and *noradrenalin*, whose effects are similar. Both help to prepare the body for emergencies by stimulating reactions that increase the supply of glucose and oxygen to the skeletal and heart muscles ("fight-or-flight" response).

The adrenal cortices produce many different steroid hormones. The cortical hormones may be grouped into three functional categories: (1) those regulating carbohydrate and protein metabolism, the *glucocorticoids*; (2) those regulating salt and water balance, the *mineralocorticoids*; and (3) those that function as sex hormones. The cortical hormones may be involved in the body's reaction to stress.

The *thyroid gland* is located just below the larynx. It has a great affinity for iodine, which is used to synthesize two thyroid hormones (TH), *thyroxin* and *triiodothyronine*. These hormones stimulate the oxidative metabolism of most tissues in the body. Hyperthyroidism—excessive TH secretion—produces an increase in the metabolic rate with high body temperature, high blood pressure, profuse perspiration, irritability, and weight loss.

Hypothyroidism—decreased TH secretion—leads to the opposite symptoms. It can be caused by dietary iodine insufficiency or by malfunction of the thyroid itself. Continued iodine deficiency causes an enlargement of the gland known as goiter. Untreated hypothyroidism in newborn children is called cretinism; such children show retarded physical, sexual, and mental development. Cretins also show abnormal protein distribution; apparently TH also plays a role in regulating protein synthesis and distribution.

The thyroid also secretes the hormone *calcitonin*, which prevents the excessive rise of calcium ions in the blood.

The *parathyroids* are four small pealike organs located on the surface of the thyroid. The parathyroid hormone (PTH) regulates the calcium-phosphate balance between the blood and other tissues; it acts primarily on the kidneys, the intestine, and the bones.

The *posterior pituitary* is connected to a part of the brain, the hypothalamus, by a stalk. It stores and releases two hormones, *oxytocin* and *vasopressin*, which are produced in the *hypothalamus* and flow along nerves in the stalk to the posterior pituitary. The hormones are released upon nervous stimulation from the hypothalamus. Oxytocin stimulates the contraction of uterine muscles. Vasopressin has two effects: it causes constriction of the arterioles with a consequent rise in blood pressure, and it stimulates the kidney tubules to reabsorb more water.

The *anterior pituitary* produces many hormones with far-reaching effects. The hormone *prolactin* stimulates milk production by the mammary glands and also participates in reproduction, osmoregulation, growth, and metabolism of carbohydrates and fats. *Growth hormone* (STH) promotes normal growth. A serious deficiency in a child results in stunted growth; an oversupply results in a giant. Growth hormone is a powerful inducer of protein anabolism and inhibits the action of insulin.

The anterior pituitary also secretes a number of hormones that help control other endocrine organs. *Thyrotrophic hormone* stimulates the thyroid, *adrenocorticotrophic hormone* (ACTH) stimulates the adrenal cortex, and the two *gonadotrophic* hormones (FSH and LH) act on the gonads. The interaction between the anterior pituitary and these glands is an example of negative feedback. For example, when the thyroxin level in the blood is low, the anterior pituitary releases thyrotrophic hormone, which stimulates the thyroid to increase production of TH. The increased TH level then inhibits the secretion of more thyrotrophic hormone by the pituitary.

The activity of the anterior pituitary is, in turn, regulated by the hypothalamus, which produces special peptide *releasing hormones*. These hormones are carried by a special blood portal system to the anterior pituitary, where they stimulate its secretory activity. The hypothalamus is the point at which information from the nervous system influences the endocrine system and is also one of the major sites of feedback from the endocrine system.

The *pineal*, a lobe in the forebrain, secretes a hormone called *melatonin*. In some lower vertebrates melatonin lightens the skin and is involved in the control of circadian rhythms. In mammals, the amount of light influences the pineal's secretion of melatonin in an inverse relationship: the more light, the less melatonin. Melatonin, in turn, inhibits the secretion of gonadotrophic hormones.

Mechanism of hormone action Since many hormones do not actually enter their target cells, but rather form weak bonds with receptor sites on the cell membrane, the *two-messenger model* of hormonal control was proposed. According to this model, the hormone acts as an extracellular first messenger; it binds to a specific receptor site on the outer membrane surface of the target cell. The binding activates *adenylate cyclase*, which catalyzes the production of an intracellular second messenger, *cAMP*, on the inner membrane surface. The increased cAMP then interacts with cytoplasmic enzyme systems to initiate the cell's response to the hormone. The presence or absence of receptors on the membrane determines whether a cell will respond to a given hormone. Moreover, a cell may respond in the same way to two or more different hormones if it has cyclase-activating receptors for each of those hormones.

A variety of other chemicals also influence cells by their effects on cAMP levels. Some work by activating adenyl cyclase; others act by influencing the enzyme *phosphodiesterase*, which breaks down cAMP.

Many hormones do not use the cAMP system. Insulin induces a rise in the concentration of *cGMP*, a related substance with effects opposite to those of cAMP. Thyroxin enters the cell and acts more directly. The steroid hormones (S) also enter the cell; there the steroid binds to a cytoplasmic receptor molecule (R), and the complex (S-R) moves to the nucleus and interacts with the genetic material. Plant hormones may function by activating important enzymes, by regulating gene activity, or by influencing protein synthesis on the ribosomes.

Prostaglandins are a group of potent substances secreted by most animal tissues; they exhibit a wide variety of effects. Some mimic the effects of hormones, apparently by stimulating or augmenting the adenylate cyclase system; others seem to inhibit this system. Still others act independently.

Hormonal control of vertebrate reproduction
Sexual reproduction in higher animals involves bringing together two *gametes*, an egg and a sperm, which then unite in the process of fertilization to form the first cell of the new individual.

Most aquatic organisms use external fertilization. The gametes are shed directly into the water and the sperm must swim to the egg. Generally, these animals release large numbers of gametes and often go through elaborate behavioral sequences to ensure that both sexes shed gametes simultaneously.

Most land animals use internal fertilization, in which the egg cells remain in the female reproductive tract until they have been fertilized by sperm inserted by the male. The sperm swim through the fluid in the female reproductive tract. Once fertilized, the egg in its fluid-filled chamber is enclosed by membranes. It is then either surrounded by a protective shell and released, or held within the female's body until embryonic development is completed.

The male gonads are the *testes*, which lie in the *scrotal sac*. Each testis has two functional components: the *seminiferous tubules*, in which the sperm are produced, and the *interstitial cells*, which secrete male sex hormone. Mature sperm move into the much-coiled *epididymis*, where they are stored and activated. During copulation the sperm move into the *vas deferens*, which conducts them to the *urethra*, which passes through the *penis* and empties to the outside. Seminal fluid from the *seminal vesicles*, the *prostate*, and the *Cowper's glands* is added to the sperm to form the *semen*. Semen provides a fluid medium for transport, for lubrication, and for protection of the sperm from acids in the female genital tract. Its sugar provides energy for the active sperm.

During embryonic development the testes begin secreting small amounts of the male sex hormone, *testosterone*, which is crucial to the differentiation of male structures. At puberty, the hypothalamus sends releasing hormone (GnRH) to the anterior pituitary, stimulating it to release *FSH* and *LH*. The LH induces the interstitial cells to produce more testosterone; this plus the FSH induces the maturation of the seminiferous tubules and causes sperm production to begin. Once testosterone reaches appreciable levels, it stimulates the development of the secondary sexual characteristics.

The female gonads are the *ovaries*, located in the lower abdominal cavity. The ovaries produce the egg cells (*oocytes*) and secrete sex hormones. Each oocyte is enclosed within a small *follicle*. During maturation, the follicle fills with fluid. When ovulation occurs, the outer wall ruptures and the oocyte and fluid are expelled into the abdominal cavity. Cilia lining the adjacent *oviduct* create a current drawing the egg into it. If sperm are present, fertilization occurs.

Each oviduct empties into the muscular *uterus*. If the egg is fertilized it becomes implanted in the uterine wall, where the embryo develops. At its lower end the uterus connects with the tubular *vagina*, which leads to the outside. The vagina is the receptacle for the penis during copulation.

Puberty in the female begins when the hypothalamus sends more GnRH to the anterior pituitary, stimulating it to release FSH and LH. These hormones cause maturation of the ovaries, which begin secreting the female sex hormones, *estrogen* and *progesterone*. Estrogen stimulates maturation of the reproductive structures and development of the secondary sexual characteristics. The changing hormonal balance triggers the onset of the menstrual cycles.

Rhythmic variations in the secretions of gonadotrophic hormones in most mammals lead to *estrous cycles*—rhythmic variations in the reproductive tract and sex urge. The reproductive cycle differs somewhat in humans: the female is receptive to the male throughout her cycle, and the thickened uterine lining is not completely reabsorbed (as in most other mammals) if no fertilization occurs; instead part of the lining is sloughed off during *menstruation*.

At the beginning of the menstrual cycle the uterine lining is thin and there are no ripe follicles. Under stimulation of GnRH from the hypothalamus, FSH is released and stimulates the maturation of the follicles, which begin to secrete estrogen. Estrogen stimulates the uterine lining to thicken. This follicular phase lasts about nine days. The high level of estrogen apparently stimulates a surge of LH from the pituitary, triggering ovulation. The LH converts the follicle into the *corpus luteum*, which continues to secrete estrogen and begins to secrete *progesterone*. Progesterone prepares the uterus to receive the embryo by activating its many glands and by inducing other chemical changes. The high level of progesterone also suppresses the growth of new follicles. If no fertilization occurs, the corpus luteum atrophies about eleven days after ovulation and progesterone secretion falls. When this happens, the thickened uterine lining can no longer be maintained, and menstruation occurs, lasting about five days. The low level of progesterone frees the hypothalamus from inhibition and allows it to stimulate the pituitary to increase FSH secretion; thus another cycle begins.

The egg cell must be fertilized within 12 hours after ovulation. The fertilized egg, or *zygote*, moves down the oviduct and, in humans, becomes implanted in the uterine wall 8–10 days after fertilization. The embryonic membranes then develop and the *placenta* is formed. Exchange of materials between the blood of the mother and that of the embryo takes place by diffusion through the placenta. The placenta soon begins to secrete a gonadotrophic hormone that maintains the corpus luteum and its secretion of progesterone, thus sustaining the pregnancy. Later the placenta secretes estrogen and progesterone directly.

In late pregnancy, estrogen secretion by the placenta increases. This causes a rise in the placenta's prostaglandin production; this increase apparently initiates the birth process. *Oxytocin* probably contributes to the induction of labor. The hormone *relaxin* aids by enlarging the birth canal.

Initiation and maintenance of lactation by mature mammary glands after birth seems to be controlled primarily by prolactin and glucocorticoids. Oxytocin from the posterior pituitary causes the milk to be ejected into the ducts of the nipple.

QUESTIONS

Testing recall

Match each of the following functions with the associated plant hormone.

a abscisic acid   d ethylene
b auxin   e florigen
c cytokinin   f gibberellin

1 induces ripening of fruit and senescence of the plant

2 stimulates cell division

3 can cause rapid elongation when applied to intact stems of dwarf plants

4 inhibits growth of lateral buds

5 inhibits seed germination

6 stimulates flowering when the photoperiod is appropriate

7 stimulates root growth; used by plant growers

8 induces seed germination

9 participates in phototropism and geotropism

10 induces changes that prepare the plant for overwintering

74 · CHAPTER 9

Match each statement below with the associated endocrine gland(s) shown in the figure. Several glands may match the same statement.

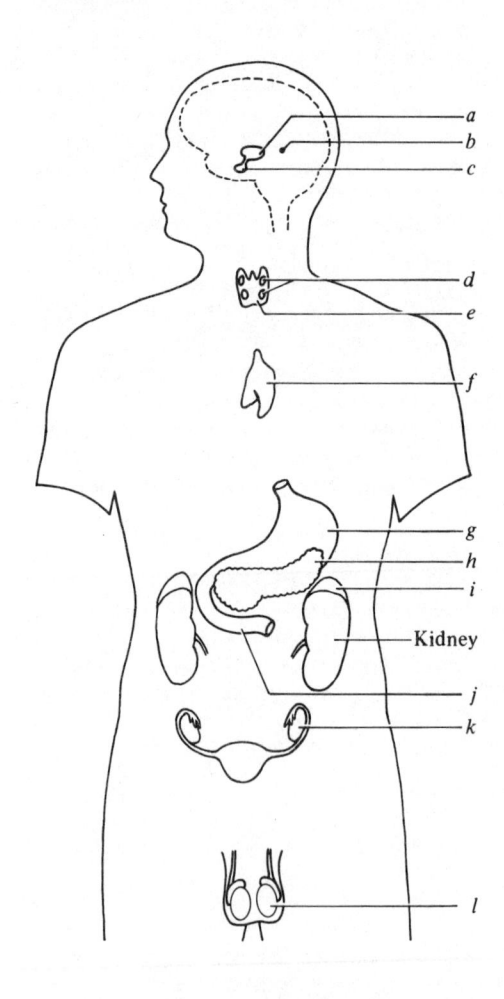

11 secretes a variety of steroid hormones

12 secretes a hormone that prevents an excessive rise in blood calcium

13 secretes the hormones that stimulate the "fight-or-flight" reaction

14 regulated by hormones from the anterior pituitary

15 produces vasopressin and oxytocin

16 produces growth hormone

17 secretes a hormone that lowers blood-glucose concentration

18 secretes a hormone that raises blood-glucose concentration

19 secretes a hormone that stimulates oxidative metabolism

20 produces a hormone that stimulates the development and maintenance of male secondary sexual characteristics

21 secretes a hormone that regulates calcium-phosphate metabolism

22 produces a hormone needed for development of the immune system

23 produces a hormone that stimulates release of bile by the gall bladder

24 secretes melatonin

25 produces releasing hormones that stimulate the anterior pituitary

Fill in the blanks.

Most mammals show rhythmic variations in the reproductive tract and sex urges called the (26) _____ cycle.

At the beginning of the menstrual cycle in humans the hormone (27) _____ from the anterior pituitary stimulates the maturation of the follicle in the ovary. The growing follicle begins to secrete the hormone (28) _____, which causes the lining of the uterus to thicken. High levels of this hormone stimulate an abrupt production of high levels of (29) _____ from the pituitary; this triggers the process of (30) _____. The old follicle is converted into the (31) _____, which secretes the hormones (32) _____ and _____. These hormones cause the uterine lining to become thicker and more glandular. High levels of the hormone (33) _____ suppress the release of (34) _____ from the hypothalamus, thereby limiting (35) _____ and _____ secretion by the pituitary.

If no fertilization occurs the corpus luteum begins to atrophy and its secretion of (36) _____ falls. Now the thickened lining of the uterus can no longer be maintained; part of it is reabsorbed and part of it is sloughed off during (37) _____.

The low levels of the sex hormones free the hypothalamus and immature follicles from inhibition, and the pituitary begins to secrete more (38) _____, beginning a new cycle.

The follicular phase of the cycle lasts about (39) _____ days, the luteal phase about (40) _____ days, and the flow phase about (41) _____ days.

If fertilization occurs, the embryo becomes implanted in the wall of the uterus about (42) _____ days after fertilization. The (43) _____ forms from the embryonic membranes and the uterine lining; here the exchange of materials between mother and embryo takes place. Part of this organ begins to secrete the hormone (44) _____ ; this preserves the corpus luteum, which continues to secrete (45) _____ . Later this organ secretes the hormones (46) _____ and _____ , which sustain the pregnancy.

Near the time of birth the secretion of the hormone (47) _____ increases. The hormone (48) _____ secreted by the posterior pituitary causes powerful uterine contractions, which aid in the birth process.

49 List, in order, all the structures through which a sperm passes from the place where it is formed to the place where fertilization occurs.

Testing knowledge and understanding

Choose the one best answer.

1 A response to changes in the length of daylight is known as

   a photoperiodism.   *d* phototaxis.
   b phototropism.   *e* photolysis.
   c photosynthesis.

2 In the experiment shown in the drawing, a transparent cap is placed on a grass shoot from which the apical meristem has been removed; also, an opaque cylinder is placed around the base. If the sun is positioned as shown, what result would you expect?

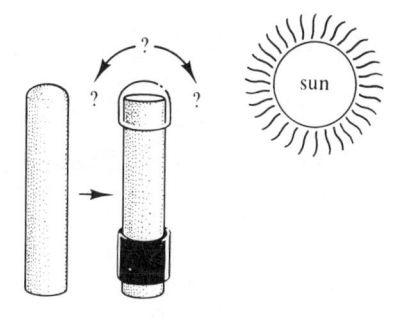

a slight movement of shoot toward sun
b pronounced movement of shoot toward sun
c slight movement of shoot away from sun
d pronounced movement of shoot away from sun
e no movement of shoot toward or away from sun

3 Plant development is often controlled by interacting hormones. In a developing tobacco stem callus, which of the combinations of cytokinin and auxin leads to the formation of buds by the callus?

76 • CHAPTER 9

4 A certain short-day plant flowers only when the night is at least 14 hours long. Under which one of the following light regimes will this plant flower?

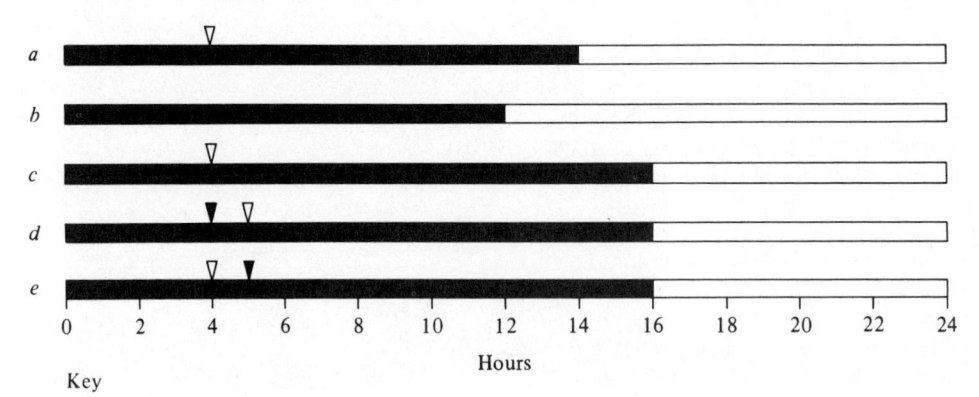

Key
White bars: daylight
Black bars: night
White triangles: intense flashes of red light
Black triangles: intense flashes of far-red light

5 An agar block, divided into X and Y halves by an impermeable partition, is inserted underneath the growing tip of an oat coleoptile. The coleoptile is then exposed for a day to light from the right, as shown in the drawing. The agar block is next removed, and the X and Y halves are placed asymmetrically atop two decapitated coleoptiles and left for two days in the dark. Which one of the following diagrams best represents the expected condition of the decapitated coleoptiles after two days?

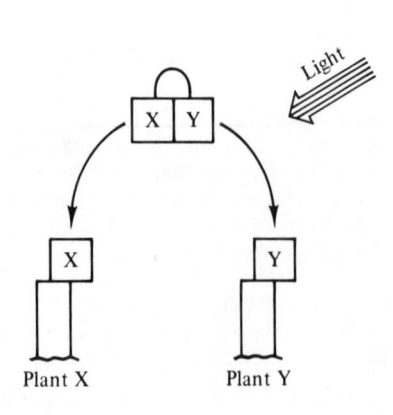

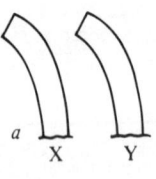

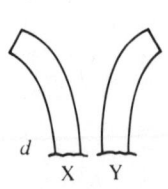

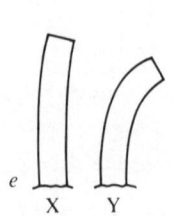

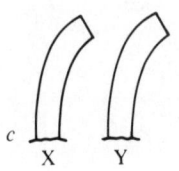

6 The procedure for demonstrating that a particular organ or tissue has an endocrine function is similar from one experiment to another. The techniques used by Banting and Best in their pioneering work with insulin have their experimental counterparts in studies on plants. The Banting and Best experiments served the same purpose as the one in plant phototropism in which

a the Darwins removed the apical meristem of a plant and exposed the plant to the sun.
b the Darwins placed an opaque cap over the tip of a plant and exposed the plant to the sun.
c Went placed a foreign apical meristem on a shoot stump in the dark.
d Went placed an apical meristem on an agar plate in the sunlight.
e Went placed an apical meristem on an agar plate in the dark.

7 The proper metamorphosis of the moth *Cecropia* (and presumably many other insects as well) involves the antagonistic interaction of two hormones. These are

a brain cell hormone and ecdysone.
b brain cell hormone and the hormone produced by the corpora allata.
c thyroxin and a pituitary hormone.
d brain cell hormone and the hormone produced by the prothoracic gland.
e juvenile hormone and ecdysone.

8 In a normal person, insulin is secreted by the pancreas immediately after ingestion of carbohydrates. Which *one* of the following is *not* an effect of insulin?

a It promotes the transport of glucose from the blood into muscle cells.
b It stimulates glycogen synthesis from glucose by the liver.
c It promotes the synthesis of fats from glucose in adipose cells.
d It stimulates the convoluted tubules of the kidney to reabsorb more sugar.
e It acts to lower blood-glucose levels.

9 In persons who have an abnormally high level of sugar in the urine, the production of insulin by the pancreas is usually

a abnormally high.   b abnormally low.

10 Which one of the following organs does not play an important role in the regulation of blood-glucose levels?

a parathyroid   d pituitary
b thyroid   e hypothalamus
c adrenal cortex

11 An iodine deficiency in the human diet is most likely to lead to

a an elevated blood calcium level.
b an elevated blood-glucose level.
c excessive inflammation reactions.
d lowered metabolic rate.
e an increase in insulin secretion.

12 An increase of thyrotrophic hormone in the blood of a mammal causes

a an increased blood supply to the thyroid.
b increased secretion of thyroxin.
c regression of the thyroid gland.
d increased nervous stimulation of the thyroid.
e reduced secretion of thyroxin.

13 A physician examines an obese patient with low blood pressure. The patient complains of feeling very lethargic and mentally dull much of the time. Of the following endocrinological malfunctions, the patient's condition is most likely to be

a diabetes.
b hypoparathyroidism.
c hyperparathyroidism.
d hypothyroidism.
e hyperthyroidism.

14 When a person takes thyroid tablets to increase his blood level of thyroid hormone,

a the secretion of TRH by the hypothalamus is inhibited.
b the secretion of thyrotrophic hormone increases.
c the thyroid gland enlarges.
d the anterior pituitary is stimulated to produce more hormones.
e the thyroid gland produces more TH.

15 A patient has an endocrine malfunction that results in excessive irritability of the muscles and nerves, which respond even to minor stimuli with tremors, cramps, and convulsions. Blood tests show an abnormally high concentration of phosphate and an abnormally low concentration of calcium. This condition is probably

a hyperthyroidism.
b hypothyroidism.
c hyperparathyroidism.
d hypoparathyroidism.
e hyperadrenocorticalism.

16 All of the following statements are true of oxytocin *except* that
 a it is transported by the blood.
 b it is produced by the hypothalamus.
 c it stimulates release of milk from the mammary glands.
 d it stimulates contraction of uterine muscles.
 e it stimulates milk production by the mammary glands.

17 The injection of a small quantity of posterior pituitary extract into the blood is likely to cause
 a an increase in the volume of urine.
 b a decrease in the volume of urine.
 c an effect similar to that of injecting a concentrated salt solution into the blood.
 d a and c.
 e b and c.

18 Which one of the following hormones is *not* secreted by the anterior pituitary?
 a FSH
 b thyrotrophic hormone
 c cortisone
 d LH
 e growth hormone

19 A person with a hypofunctioning anterior pituitary would probably show all of the following symptoms except
 a decreased metabolic rate.
 b decreased activity of the adrenal cortex.
 c sexual immaturity.
 d increased urine output.
 e decreased growth hormone secretion.

20 According to the proposed two-messenger model for the mode of action of animal hormones,
 a the hormone enters the target cell by combining with a carrier molecule and moves to the nucleus, where it reacts with genetic material.
 b the hormone enters the target cell, where it causes production of prostaglandins which then interact with the genetic material in the nucleus.
 c the hormone interacts with a specific receptor site on the outer surface of the cell so as to influence the concentration of cyclic AMP within the cell.
 d the first hormone is produced by the pituitary and it in turn causes the secretion of a second hormone from another endocrine gland.

21 Which one of the following hormones requires the involvement of cAMP in order to bring about a change in the target cells?
 a mineralocorticoids   d thyroxin
 b sex hormones   e adrenalin
 c all plant hormones

22 Which one of the following animal's eggs lacks an amnion?
 a human   d black snake
 b cow   e codfish
 c robin

23 The prostate gland plays a major role in the production of
 a seminal fluid.   d urine.
 b sperm.   e amniotic fluid.
 c testosterone.

24 The role of FSH in humans is to
 a stimulate the development of the follicle in the ovary and production of sperm in the testes.
 b stimulate the growth of the corpus luteum in the ovary and production of the sperm in the male.
 c stimulate the interstitial cells of the male testes.
 d stimulate the menstrual flow.
 e stimulate uterine contractions at birth.

25 In the menstrual cycle, blood progesterone levels are highest during
 a the follicular (growth) phase.
 b the time of ovulation.
 c the luteal (secretory) phase.
 d the first part of the flow phase.
 e the last part of the flow phase.

26 Which is the correct sequence of organs involved in producing progesterone?
 a hypothalamus–anterior pituitary–ovary
 b hypothalamus–posterior pituitary–ovary
 c anterior pituitary–adrenal cortex–uterus
 d posterior pituitary–adrenal medulla–ovary
 e hypothalamus–ovary–uterus

27 Which one of the following statements is *not* true?
 a Most birth-control pills are mixtures of estrogen and progesterone (or close analogs) designed to inhibit the hypothalamus.
 b Vasectomy does not alter a man's endocrine system, and hence should not reduce either sexual drive or sexual competency.
 c Tubal ligation does not alter a woman's endocrine system, and hence should not

reduce either sexual drive or sexual responsiveness.

   d The diaphragm is a birth-control device that, when properly fitted, covers the cervix and prevents sperm from entering the uterus.

   e In the rhythm method of birth control, sexual abstinence on days 14–16 in each menstrual cycle reliably prevents pregnancy.

28 The birth-control pills most commonly used in the United States today are mixtures of estrogen-like and progesterone-like compounds. They probably prevent pregnancy mainly by

   a stimulating the release of FSH from the anterior pituitary.
   b preventing the glands of the uterine lining from becoming secretory.
   c decreasing the mobility of the sperm by inhibiting uterine contractions.
   d blocking secretion of gonadotrophic-releasing hormone by the hypothalamus.
   e causing premature release of the egg from the follicle.

29 Which one of the following functions does the placenta *not* perform?

   a removes waste materials from the fetus
   b supplies oxygen to the fetus
   c secretes hormones
   d provides the fetus with nutrients
   e replaces used fetal erythrocytes with new ones from the mother

30 If gonadotrophins (HCG, LH) from human placenta were injected into a mature male, one might expect to see

   a an increase in estrogen and progesterone.
   b an increase in testosterone.
   c a decrease in the activity of the testes.
   d a 28-day cycling of male hormones.
   e increased sperm production by the testes.

For further thought

1 Explain the role of the hormones auxin, gibberellin, abscisic acid, and cytokinins in the germination of the seed and growth of the young plant.

2 Describe and explain the function of the abscission layer in the stem of the fruit or leaf, and discuss the role of hormones in controlling its development. How can fruit drop be prevented?

3 The poinsettia is a short-day plant with a critical night length of 12 hours. How could you get the plant to flower in June?

4 List the hormones that help control carbohydrate metabolism and protein metabolism.

5 Discuss the relationship between the hypothalamus and the anterior and posterior pituitary. Describe the role of the anterior pituitary and hypothalamus in coordinating the activity of the thyroid gland.

6 At the age of nine years a girl was taken to a doctor who suspected that she had some sort of endocrine deficiency. At that time she was only 36 inches tall, although correctly proportioned. By 17, she was only as tall as a nine-year-old. There was no sexual development during her adolescent years.

   a What type of endocrine deficiency might this child have?
   b What hormones is she not producing?
   c What other symptoms might she have?
   d What hormonal therapy might be prescribed?

Chapter 10

NERVOUS CONTROL

KEY CONCEPTS

1 The evolution of nerve and muscle tissues with their capacity for rapid response was basic to the evolution of active multicellular animals.

2 The nervous system is composed of cells that are specialized to detect changes in the environment and to conduct information to the effectors, which produce a rapid response.

3 Increasing the number of conducting cells in the nervous pathway, by making possible more alternative routes, increases the flexibility of the response to a stimulus.

4 Animals with more advanced nervous systems have evolved more complex pathways, a greater degree of centralization, and better developed sense organs than animals with simple nervous systems.

5 The reflex arc is the basic functional unit of the nervous system.

6 Nerve impulses are conducted along the neuron by electrochemical changes; they are transmitted across the synapse by chemicals.

7 Each type of sensory receptor functions as a transducer, converting the energy that constitutes the particular stimulus to which it is attuned into the electrochemical energy of the nerve impulse.

8 The sensation perceived is determined by the part of the brain stimulated; it is not inherent in the stimulus or in the neural message.

9 Light detection is similar in all animals that respond to light; light energy produces changes in a light-sensitive pigment.

10 The major evolutionary change in the vertebrate brain was the steady increase in the size and importance of the cerebrum, with a corresponding decrease in the relative size and importance of the midbrain.

11 In the human brain, the cerebral cortex has taken over many of the functions of other parts of the brain and has become dominant.

OBJECTIVES

After studying this chapter and reflecting on it, you should be able to

1 Define irritability and name its three principal components.

2 Describe the structure of a typical neuron and differentiate among sensory, motor, and interneurons.

3 Contrast the nervous system of radially symmetrical animals with that of bilaterally symmetrical animals.

4 List the six major evolutionary trends in bilateral nervous systems and explain how these trends are exemplified in flatworms.

5 Contrast the nervous system of vertebrates with that of the annelids and arthropods with respect to the location, form, and structure of the longitudinal nerve cord(s); the degree of centralization; and the dominance of the brain.

6 Diagram a spinal reflex arc and relate it to the structure of a spinal nerve and the spinal cord.

7 Differentiate among the central, somatic, and autonomic nervous systems.

8 Contrast the function of the sympathetic and parasympathetic nervous systems.

9 Explain how the somatic reflex control of breathing differs from the autonomic control of the rate of heartbeat.

10 Explain in some detail how an impulse is transmitted along a neuron. Include in your explanation the terms threshold, action potential, and all-or-none response. Discuss the roles of the sodium-potassium pump and myelin sheath (when present).

11 Explain how the normal resting potential of the neuron is maintained.

12 Draw a typical synapse and discuss the events of synaptic transmission.

13 List the main transmitter chemicals and indicate whether they act at synapses within the central nervous system or outside the central nervous system.

14 Explain the action of transmitter substances on the postsynaptic membrane, and explain the difference between an excitatory and inhibitory synapse.

15 Describe how neurological drugs affect transmission across the synapse.

16 Contrast impulse transmission across the junction between nerve and muscle with transmission across a synapse between neurons.

17 Explain why the sympathetic nervous system and the hormones of the adrenal medulla have similar effects on the body.

18 Give three examples of the close relationship between the nervous system and the endocrine system.

19 Describe the different types of sensory receptors in the skin, skeletal muscles, and viscera, and explain how a stimulus can lead to the production of a generator potential in the various types of receptor cells.

20 Explain the statement that sensation is a creation of the brain, not a property of the specialized receptors or the stimulus.

21 Contrast the receptors for taste and smell with respect to location, basic function, type of receptor cell, and possible mode of action.

22 Distinguish between a camera-type eye and a compound eye.

23 Describe, using diagrams, the structure and function of the human eye. Contrast the structure, location, and function of the two principal types of visual receptor cells.

24 Discuss the processes of refraction and accommodation in the human eye.

25 Describe, using diagrams, the structure and function of the human ear.

26 Explain how the inner ear determines the head's position with respect to gravity, and how it detects changes in position.

27 Describe the sensory hair cells of the lateral-line system and relate their structure to the receptors for hearing and equilibrium.

28 Describe the evolutionary history of the vertebrate brain from fish through amphibians and reptiles to mammals.

29 Discuss the role of the thalamus and the reticular system.

30 List three functions of the hypothalamus.

31 Discuss the general structure and functions of the human cerebral cortex.

SUMMARY

Because hormonal control is slow, the evolution of nerve and muscle tissue was basic to the evolution of multicellular animals and their active way of life.

All protoplasm has *irritability*—the capacity to respond to stimuli. Irritability involves three components: reception of the stimulus, conduction of a signal, and response by an effector. All three components are carried on within the cell protoplasm in unicellular organisms; some possess specialized structures for these processes. All multicellular animals (except sponges) have evolved some form of nervous system. Most nervous pathways have at least three separate cells: receptor, conductor, and effector cells. More complex pathways have additional conductor cells; these increase the flexibility of response.

The typical nerve cell, or *neuron*, consists of the *cell body*, which contains the nucleus, and one or more long *nerve fibers* that extend from it. *Sensory neurons* lead from receptor cells, *motor neurons* lead to effector cells, and *interneurons* lie between the sensory and motor neurons. Junctions between neurons are called *synapses*.

Radially symmetrical animals, such as coelenterates, have a simple nervous system without a brain. We can see six major trends in the evolution of the nervous system of bilaterally symmetrical animals:

1 increased centralization by the formation of longitudinal nerve cords;
2 increased complexity of pathways;
3 the formation of distinct functional areas and structures;
4 increased cephalization;
5 one-way conduction;
6 increased number and complexity of sense organs.

These evolutionary trends are most developed in the vertebrates and in the annelids and arthropods. The main difference between the central nervous system (brain and spinal cord) of vertebrates and those of annelids and arthropods is that vertebrates have a single dorsal, hollow nerve cord whereas annelids and arthropods have two ventral, solid nerve cords. The vertebrate brain is also more highly developed and exerts more dominance over the entire nervous system.

Nervous pathways in vertebrates Vertebrate neurons usually have two types of fibers: *dendrites*, which conduct impulses to the cell body, and *axons*, which conduct impulses away from the cell body. Within the central nervous system the neurons are associated with vast numbers of *neuroglia* cells. The function of many of the neuroglia is uncertain, but some are known to wrap around and around axons, forming the *myelin sheath*. Outside the central nervous system, many vertebrate axons are enveloped by *Schwann cells*, which may also form a myelin sheath. Myelin sheaths speed up the rate of conduction.

A *reflex arc* is a simple neural pathway linking a receptor and effector. Most somatic reflex arcs begin with a *sensory neuron* that conducts the impulse from the receptor to the dorsal portion of the spinal cord, where the sensory neuron synapses with *interneurons*. These in turn synapse with *motor neurons* in the cord, and the impulses are conducted along their axons to the effectors (usually skeletal muscles), which respond to the stimulus. Reflex arcs always interconnect with other neural pathways; the interneurons connect with pathways leading to the brain, and the brain can send impulses that modify the reflexes.

A *nerve* consists of a number of neuron fibers bound together. Some nerves have only sensory fibers and are therefore *sensory nerves*, others are purely *motor*, and still others have both types of fibers and are *mixed nerves*.

The *autonomic nervous system* (ANS) consists of nervous pathways that conduct impulses from the *central nervous system* (CNS) to various internal organs. The autonomic nervous system regulates the body's involuntary activities. Autonomic pathways usually have two motor neurons. The first neuron exits from the central nervous system and synapses with a second that innervates the target organ. The autonomic nervous system is separated into two parts, *sympathetic* and *parasympathetic* systems. Most internal organs are innervated by both, with the two systems usually functioning in opposition to each other.

Transmission of nervous impulses A nerve impulse is a wave of electrochemical change moving along a nerve fiber. The potential stimulus must be above a critical intensity, or *threshold* value, to initiate an impulse. If the axon fires, it will fire maximally or not at all—an *all-or-none response*.

The inside of a resting nerve fiber is negative relative to the outside because the ratio of negative to positive ions is higher inside the cell than outside. The inside has a high concentration of potassium ions (K^+) and negative organic ions; the outside has a high concentration of sodium ions (Na^+). When a fiber is stimulated the membrane becomes permeable to Na^+ ions, which cross the membrane into the cell, making the inside positively charged relative to the outside. An instant later, the membrane becomes permeable to K^+ ions, which rush out of the cell, restoring the original charge. This cycle of electrical changes is known as the *action potential*. The action potential at the point of stimulation alters the permeability at adjacent points and initiates the same cycles of changes there. Active transport restores the original ion distribution (high K^+ inside, high Na^+ outside).

The polarization of the nerve cell membrane depends on three interacting factors: the *sodium-potassium pump*, the greater leakage of K^+ ions than Na^+ ions, and the inability of the negative organic ions to leave the cell.

Transmission across synapses The axon of one neuron usually synapses with the dendrites or cell bodies of other neurons. Each tiny branch of an axon usually terminates in a *synaptic bouton*. A few synapses are electrical; in these the bouton and adjoining cell membrane are connected by a gap junction, permitting direct electrical transmission from the first neuron to the second. Most synapses, however, are chemical. When an impulse traveling along the axon reaches the bouton it causes the *synaptic vesicles* to discharge their stored *transmitter* chemical into the synaptic cleft. The transmitter molecules diffuse across the cleft and alter the membrane potential of the next neuron. Synaptic transmission is slower than impulse conduction along the neuron. It is the chemical synapses that make transmission along the neural pathways one-way.

The transmitter chemical in the peripheral nervous system is *acetylcholine*. After acetylcholine has exerted its effect on the postsynaptic neuron it must be destroyed by the enzyme *cholinesterase*. Transmitter chemicals inside the central nervous system include *acetylcholine*, *noradrenalin*, *serotonin*, *dopamine*, and *GABA*. Transmitter chemicals can be excitatory or inhibitory. An *excitatory* transmitter slightly reduces the polarization of the postsynaptic membrane, creating an excitatory postsynaptic potential (*EPSP*). If the EPSP reaches threshold, it triggers an impulse. An *inhibitory* transmitter increases the polarization of the postsynaptic membrane, a condition called an inhibitory postsynaptic potential (*IPSP*). This makes the neuron harder to fire. Synaptic transmitter substances may act by binding to receptor proteins in the membrane, thus opening channels through which ions can pass.

Some synapses are also subject to *presynaptic inhibition*; in this case inhibitor neurons impinging on the bouton reduce the number of vesicles that can release transmitter.

Synapses are points of resistance in nervous pathways; enough excitatory transmitter must be released within a short time to build up sufficient EPSP to initiate an impulse. The cell integrates all the excitatory and inhibitory signals it receives, and either fires or remains silent.

Synapses are the regulatory valves of the nervous system; they determine the routes impulses will follow through the nervous system. Their operation depends on a delicate balance between transmitter substance, deactivating enzyme, and membrane sensitivity. Synaptic malfunctions have been implicated in several mental disorders. Neurological drugs can alter synaptic function in a variety of ways.

The gap between the axon and the muscle it innervates is called the *neuromuscular junction*. Transmission across this gap is also via transmitter chemicals. Acetylcholine is the transmitter at neuromuscular junctions in both the somatic and parasympathetic systems. Noradrenalin is the transmitter in the sympathetic system. This explains the similar effects of the sympathetic nervous system and the hormones of the adrenal medulla; both release noradrenalin.

Sensory reception Specialized receptor cells are the body's principal means of gaining information about its environment; they function as transducers, converting the energy of a stimulus into the electrochemical energy of a nerve impulse. Each type of receptor is responsive to a particular kind of stimulus.

Stimulation of a sensory receptor produces a local depolarization (*generator potential*) of the receptor cell. When the generator potential reaches threshold level, it triggers an action potential in the nerve fiber.

Each receptor sends impulses to a particular part of the brain. It is the part of the brain to which the impulses go, not the stimulus, the receptor, or the message itself, that determines the quality and location of the sensation.

The skin contains sensory receptors for touch, pressure, heat, cold, and pain. These receive information from the outside environment. Receptors inside the body receive information about the condition of the body itself.

The receptors of taste and smell are chemoreceptors; they are sensitive to solutions of different kinds of chemicals, which can bind to them by weak bonds. The receptors for the four taste senses are located in *taste buds* on different areas of the tongue. The sensations we experience are produced by a blending of these four basic sensations. The receptor cells for smell are located in the upper part of the nasal passages. These are specialized to detect vapors from distant sources; they are much more sensitive than taste receptors.

Vision Almost all animals respond to light stimuli. Most multicellular animals have evolved specialized light receptor cells containing a pigment that undergoes a chemical change when exposed to light.

The light receptors of many invertebrates simply detect the presence of light, and perhaps differences in intensity. Receptor organs capable of detecting the direction of a light source frequently contain many sensory cells oriented at different angles. More complex eyes usually include a lens capable of concentrating light on the receptor cells. Lenses made possible the evolution of image-forming eyes.

There are two basic types of image-forming eyes: *camera-type eyes* (in some molluscs and vertebrates) and *compound eyes* (in insects and crustaceans). The camera-type eye uses a single-lens system to focus light on the many receptor cells that make up the *retina*. A compound eye is made up of many closely packed functional units called ommatidia, each of which acts as a separate receptor. Image formation depends on the light pattern falling on the surface of the compound eye. Since each ommatidium points in a slightly different direction, each is stimulated by light coming from different points.

Structure of the human eye The light rays coming into the eye are focused by the *cornea* and *lens* on the light-sensitive *retina*, which contains the specialized receptor cells, the *rods* and *cones*. The sensitive rod cells function in dim light, the cones in bright light. The cones enable us to detect color.

The light-sensitive pigment in the rods (*rhodopsin*) is converted into a different form when struck by light, and is regenerated in the dark. The pigment conversion leads to a change in membrane permeability and an impulse is generated. Cone vision is more complex; apparently there are three types of cones, each containing a different pigment and each sensitive to different wavelengths of light.

Well-defined image vision depends on precise focusing of the incoming light on the retina. The shape of the lens is alterable; it makes possible adjustments in the focus (*accommodation*) depending on whether the object being viewed is close or distant.

Hearing Receptors of the sense of hearing are specialized for detection of vibrations. In humans, vibrations in the air pass down the *auditory canal* of the outer ear and strike the *tympanic membrane*, causing it to vibrate. The vibrations are increased in force as they are transmitted by three small bones across the middle ear to the *oval window*. The resultant movement of the oval window produces movement of the fluid in the canals of the *cochlea*, causing the *basilar membrane* to move up and down and rub the hair cells of the *organ of Corti* against the overlying *tectorial membrane*. The stimulus to the hair cells is passed into the associated sensory neurons, which carry impulses to the auditory centers of the brain. Humans can distinguish between three characteristics of sounds: pitch, volume, and intensity.

The upper portion of the inner ear consists

of three *semicircular canals* and a large *vestibule* that connects them to the cochlea. The sensory hair cells lining the two cavities of the vestibule send information to the brain about the position of the head relative to gravity. The three semicircular canals are concerned with sensing changes in the speed or direction of movement.

Fish have an important sensory system called the *lateral-line system*, which consists of clusters of hair cells in grooves along the head and sides. Movements of water bend the hairs, enabling the fish to measure its progress and to detect water disturbances created by other objects. It is believed that the sensory-hair apparatus of hearing and equilibrium in terrestrial vertebrates evolved from the lateral-line system.

The brain The brains of invertebrate animals are much smaller in relation to the size of their bodies than those of vertebrates, and their dominance over the rest of the nervous system is usually less. The brains of primitive vertebrates consist of three irregular swellings at the anterior end of the nerve cord. These three regions, the *forebrain*, *midbrain*, and *hindbrain*, have been much modified in the evolution of the more advanced vertebrates.

Very early in its evolution the forebrain was divided into the *cerebrum* and the more posterior *thalamus* and *hypothalamus*. The midbrain became specialized as the *optic lobes*, and the hindbrain was modified to form the ventral *medulla oblongata* and the *cerebellum*. The hindbrain has changed little over the course of evolution; the medulla continues to be a control center for some autonomic and nervous pathways, and the cerebellum is still concerned with equilibrium and muscular coordination. The major evolutionary change has been the enormous increase in the size and relative importance of the cerebrum, with a corresponding decrease in the midbrain.

In certain advanced reptiles, a new area of the cerebral cortex, the *neocortex*, evolved. In mammals the neocortex expanded to cover the surface and dominate the other parts of the brain.

The mammalian forebrain The thalamus contains part of the important *reticular system*, a network of neurons that runs through the medulla, midbrain, and thalamus. The reticular system receives inputs from "wiretaps" on all incoming and outgoing brain communication channels. It activates the brain upon receipt of stimuli and is an indispensable filter that lets only a few of the major sensory inputs reach the brain's higher centers.

Besides serving as a crucial link between the nervous and endocrine systems, the *hypothalamus* is also the control center for the visceral functions of the body and a major integrating region for emotional responses.

The proportion of the cerebral cortex taken up by purely motor and sensory areas is smaller in humans than in other animals; the association areas occupy the greatest proportion of the cortex. The area of the cortex devoted to each body part is proportional to the importance of that part's sensory or motor activities. Some association areas have been mapped; areas concerned with speech and memory have been studied.

Memory apparently involves the potentiation of a neuronal circuit by repeated use of that circuit. Stimulation may cause chemical changes in the neurons. It has been found that increases in the amount of RNA and protein synthesis may be involved in storing a memory.

QUESTIONS

Testing recall

Complete the following statements by underlining the correct term or terms.

1 The components of irritability are (sensory reception, conduction, response).

2 Increasing the number of conducting cells in a nervous circuit (increases, decreases) the flexibility of response.

3 The central nervous system consists of the (brain, spinal cord, autonomic nerves, spinal nerves).

4 Radially symmetrical animals such as coelenterates have a nervous system that is (highly centralized, a diffuse nerve net).

5 The nervous system of the advanced flatworms is characterized by (nerve net, two ventral nerve cords, a dorsal nerve cord, a highly developed brain, simple sense organs at the anterior end).

6 The nerve cord(s) of an insect is (are) (single, double), (solid, hollow), and

(dorsal, ventral). (Many, few) ganglia are present. The brain exerts (limited, extensive) dominance over the entire nervous system.

7   In the vertebrate neuron, the (axons, dendrites) conduct impulses toward the cell body and the (axons, dendrites) conduct impulses away from the cell body. Each neuron usually has (one, many) dendrite(s) and (one, many) axon(s). (Sensory neurons, motor neurons, interneurons) conduct impulses to the central nervous system; (sensory neurons, motor neurons, interneurons) conduct impulses within the central nervous system.

8   Within the central nervous system special satellite cells called (Schwann cells, neuroglia) are associated with the neurons; frequently these wrap around and around the neuron, forming the (myelin sheath, axon, nerve).

9   The nerves of the autonomic nervous system innervate the (skeletal muscles, blood vessels, digestive tract, respiratory system, reproductive system). The autonomic nerve pathways usually have (one, two, three, many) motor neuron(s). The parts of the autonomic nervous system are the (somatic, sympathetic, parasympathetic) divisions.

10 Breathing is controlled by (somatic reflex arcs, the autonomic nervous system). The stimulus for breathing is the concentration of (CO_2, oxygen) in the blood.

11 The heart rate is (accelerated, slowed) by the sympathetic nervous system. (Low O_2, high CO_2, high blood pressure) act(s) to speed up the heart rate.

12 The resting nerve fiber is polarized with the inside (positive, negative) compared to the outside. When a stimulus is above threshold, the membrane first becomes permeable to (Na^+, K^+, organic) ions, which rush (into, out of) the cell; the inside of the cell is now charged (positively, negatively) with respect to the outside. Next the membrane becomes permeable to (Na^+, K^+, organic) ions and these rush (into, out of) the cell, restoring the original polarization.

13 The original ionic balance is restored by (the sodium-potassium pump, active transport, facilitated diffusion).

14 Myelinated fibers conduct impulses (slower, faster) than unmyelinated fibers.

15 Most synapses are (electrical, chemical). When an impulse traveling down the axon reaches the bouton, it makes the membrane of the bouton permeable to (Na^+, K^+, Ca^{++}), which diffuses into the bouton and promotes the release of (Na^+, K^+, transmitter chemicals) into the synaptic cleft. Synaptic transmission is much (slower, faster) than impulse conduction along the axon.

16 The transmitter substance at neuron-neuron synapses outside the central nervous system is generally (noradrenalin, serotonin, acetylcholine).

17 An excitatory transmitter substance slightly (reduces, increases) the polarization of the postsynaptic membrane, making it (easier, more difficult) to trigger an impulse. If the transmitter had been inhibitory, the membrane would have become (depolarized, hyperpolarized), a condition called (IPSP, EPSP).

18 If both excitatory and inhibitory transmitters converge on a cell at the same time, their effects are combined in the process called (summation, integration).

19 The gap between the end of the motor axon and the effector it innervates is called the (motor end plate, neuromuscular junction). The transmitter substance released by the motor neurons at their junction with the effectors in both somatic and parasympathetic neurons is (acetylcholine, noradrenalin), whereas the second motor neuron of the sympathetic pathways releases (acetylcholine, noradrenalin).

20 Sensory receptors may be (portions of nerve cells, specialized cells); most are responsive to (one, many) kind(s) of stimulus.

21 The sensation experienced depends on the (type of stimulus, neural message, part of the brain stimulated).

22 When a sensory receptor is stimulated, a small local depolarization called the (action potential, generator potential) is produced. When this reaches threshold level, it triggers a(n) (action potential, generator potential) in the nerve fiber. The (generator potential, action potential) increases in direct relation to the increase in intensity of the stimulus.

23 Taste and smell receptors are (touch receptors, chemoreceptors). Taste receptors

are (neurons, specialized receptor cells), whereas olfactory receptors are (neurons, specialized receptor cells).

24 (Compound eyes, camera-type eyes) are capable of forming images. Camera-type eyes are found in (molluscs, insects, crustaceans, vertebrates).

25 Light entering the human eye must first pass through the transparent (sclera, cornea, iris); it then passes through an opening in the (sclera, iris, retina). When the circular muscles of the iris contract, the size of the pupil is (enlarged, reduced).

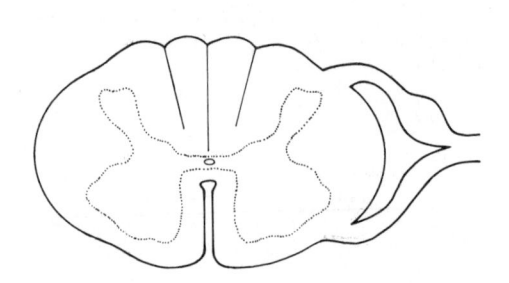

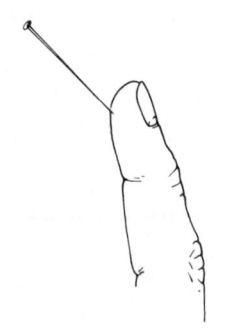

26 The rod cells are more numerous in the (center, periphery) of the retina and function in (dim, bright) light. They are responsible for (color, black and white) vision.

27 Within the retina there is (are) (one, several) set(s) of synapses that enable the eye to modify information.

28 Focusing the light rays on the retina is accomplished by the (lens, cornea). When viewing distant objects the smooth muscles of the ciliary body are (contracted, relaxed), the tension on the suspensory ligament is (high, relaxed), and the lens is (thicker, thinner).

29 Low-frequency sounds stimulate the hair cells near the (base, tip) of the cochlea. Tone quality is detected when hair cells in (one, several) region(s) of the cochlea are stimulated at the same time.

30 The hair cells of the semicircular canals function in sensing (the position of the head, acceleration).

31 Below are listed nine parts of the human body involved in hearing. Arrange them in the order in which they function in sound reception.

a auditory centers in the brain
b auditory canal
c cochlear fluid
d middle ear bones
e organ of Corti
f oval window
g pinna
h sensory neurons
i tympanic membrane

32 Suppose you pricked your finger with a pin. On the accompanying diagram, draw in and label the components of a reflex arc that originates in the finger. Include the sensory neuron, interneurons, motor neuron, receptor, and effector. Label the parts of each neuron (axon, dendrites, cell body).

Matching: Match each part of the brain with its function or description.

a cerebellum
b cerebrum
c hypothalamus
d medulla oblongata
e optic lobes
f thalamus
g reticular system

33 parts of the forebrain

34 parts of the midbrain

35 parts of the hindbrain

36 a major sensory-integration center in lower vertebrates

37 filter for incoming sensory information

38 area concerned with balance, equilibrium, and muscular coordination

39 important in regulating the endocrine system

40 site of somatic sensory area

41 vital centers for breathing and heart rate

42 associated with memory and learning

43 centers for thirst and hunger

44 site of centers for smell

Testing knowledge and understanding

Choose the one best answer.

1 Animal control centers tend to be concentrated in the head region; this is known as

   a autonomic control. c cephalization.
   b corticalization.   d rationalization.

2 Another evolutionary development that paralleled the above tendency in the vertebrates was

   a the increased size and length of the spinal cord.
   b the concentration of major sense organs in the head region.
   c the increased coordinating ability of the endocrine system.
   d the development of efficient locomotor appendages in higher animals.
   e the development of the nerve net.

3 The diagram below depicts a typical vertebrate reflex arc involving a single receptor (R), a muscle (M), and three neurons. Assume that you apply an electrical stimulus at point X. What is the *most distant* point from X at which you would expect to detect a signal?

4 A certain neuron is encountered during dissection of the muscles in the leg of a pig. It is part of the voluntary nervous system. Which of the following observations indicates that it is a motor neuron, not a sensory neuron?

   a Its cell body is located inside the spinal cord.
   b It exhibits an all-or-none response to stimuli.
   c The end of its axon secretes a neurotransmitter.
   d Its axon is myelinated.

5 Which statement below most closely fits the knee-jerk reflex arc?

   a It involves more than two synaptic junctions, but no brain control.
   b It involves one synaptic junction and no brain control.
   c It normally involves two synaptic junctions and can be modified by the brain.
   d It normally involves one synaptic junction and can be modified by the brain.
   e The stretch receptor connects with the brain stem and sends impulses to the dorsal root ganglion motor neurons.

6 Which of the following is *not* a function of the autonomic nervous system?

   a causing an endocrine gland to secrete hormones
   b contracting certain blood vessels and dilating others
   c causing changes in the heart rate
   d causing the muscles of the intestinal wall to contract
   e causing the muscles of the arm to contract in response to a pinprick

7 Which one of the following is *not* a function of the sympathetic nervous system?

   a dilation of intestinal capillaries
   b acceleration of the rate of heartbeat
   c dilation of capillaries in some skeletal muscle
   d erection of hairs on the skin
   e decrease of peristalsis of the digestive tract

8 The rate of the human heartbeat is controlled by which of the following? (Pick the most complete and specific answer.)

   a sympathetic nervous system
   b parasympathetic nervous system
   c somatic nervous system
   d autonomic nervous system
   e peripheral nervous system

For question 11

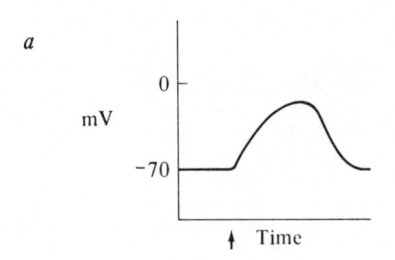

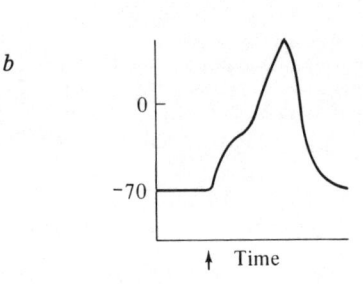

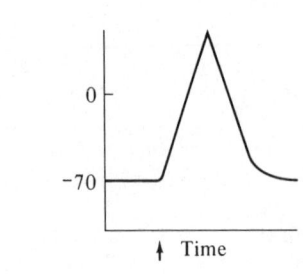

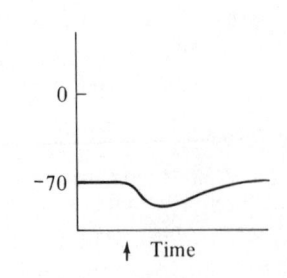

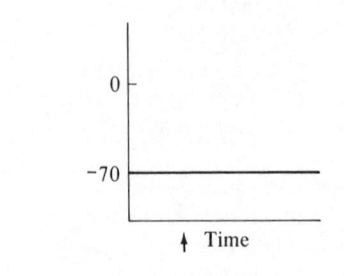

a

b

c

d

e

9 Which of the following is *not* true of the axon in a resting (nonconducting) stage?
   a The membrane is relatively impermeable to sodium ions.
   b The membrane has a positive charge on the outside and a negative charge on the inside.
   c The membrane is highly permeable to large negatively charged organic ions and allows them to leak out.
   d Potassium ions are in higher concentration inside the axon than outside.
   e Sodium ions are in higher concentration outside the axon than inside.

10 Of the following activities, which is the *second* event to occur in the depolarization of a nerve cell?
   a Na^+ rushes inside.
   b Na^+ rushes outside.
   c K^+ rushes inside.
   d K^+ rushes outside.
   e Negatively charged ions rush outside.

11 Suppose an intracellular electrode records a resting potential for a given neuron of −70 milivolts. Which of the recordings at left best portrays what will happen if an inhibitory neurotransmitter substance is applied to the neuron? (The arrows indicate time of application.)

12 You have set up two electrical recording instruments that will record any electrical changes that occur. One is placed at each end of a single neuron 10 cm long. If at the first pair of electrodes the depolarization of the action potential is recorded at 100 mV then what will be the magnitude of the depolarization by the time the action potential reaches the second pair of electrodes at the other end?

   a 10 mV
   b less than 100 mV but cannot be more exact
   c 100 mV
   d more than 100 mV but cannot be more exact
   e 1000 mV

13 A certain neuron in a cat is located entirely outside the central nervous system. The synaptic vesicles of its axon release a transmitter substance that is not destroyed by cholinesterase. This neuron is

a a sensory neuron of the somatic (voluntary) portion of the nervous system.
b a first motor neuron of the sympathetic system.
c a second motor neuron of the sympathetic system.
d a first motor neuron of the parasympathetic system.
e a second motor neuron of the parasympathetic system.

14 Many poisons and drugs act at synapses. One kind of snake venom binds irreversibly with acetylcholine and thus, in an acetylcholine synapse, would be expected to

a prevent the postsynaptic membrane from depolarizing.
b cause the postsynaptic membrane to depolarize only once.
c cause the postsynaptic membrane to depolarize many times in succession.
d produce muscular spasms.
e prevent cholinesterase production.

15 A drug that produces which one of the following effects would probably be *least* effective as a tranquilizer?

a interferes with uptake of noradrenalin into synaptic vesicles
b interferes with synthesis of noradrenalin in nerve cells
c prevents release of noradrenalin from synaptic vesicles
d blocks receptor sites for noradrenalin on postsynaptic membranes
e enhances the sensitivity of neurons to noradrenalin

16 Which one of the following groups of structures are all derived from the forebrain?

a cerebral cortex, thalamus, cerebellum
b hypothalamus, cerebellum, medulla
c midbrain, cerebrum, hypothalamus
d thalamus, hypothalamus, cerebral cortex
e cerebral cortex, medulla, cerebellum

17 Over the course of vertebrate evolution the position of the chief control center has shifted from the

a hindbrain to the midbrain.
b midbrain to the forebrain.
c cerebellum to the cerebrum.
d cerebrum to the cerebellum.
e hindbrain to the cerebrum.

For further thought

1 As you are crossing a street, a car suddenly swerves towards you. Explain how the nervous system enables you to meet this emergency.

2 Certain drugs act as metabolic poisons and prevent the synthesis of ATP. Suppose you treated a neuron with one of these drugs. Would the neuron conduct an impulse? Explain.

3 Cocaine is a drug that inhibits the active reuptake of noradrenalin into the brain's presynaptic neurons. Would you classify cocaine as a stimulant or depressant? What effect might cocaine have on the heart rate, blood pressure, and mental awareness?

4 Distinguish between the sense of smell and the sense of taste. How does the sense of smell contribute to the sense of taste?

5 When you walk into a dark room from bright sunlight, you cannot see anything. However, after some time your vision returns. Explain the initial lack of vision and why the vision returns.

6 As people grow older, they tend to become more farsighted and often need glasses for reading. Explain why this process occurs, and how the glasses correct the problem.

7 A person in an automobile accident sustained severe head injuries. A brain scan showed damage to the central portion of the cerebrum on the right side. What symptoms would you expect this person to show?

Chapter 11

EFFECTORS

KEY CONCEPTS

1. The underlying mechanism of effectors of motion depends on either microfilaments or microtubules.

2. The principal effectors of movement in higher animals are the muscle cells—elongate cells specialized for contraction.

3. A skeleton plays an important role in effecting movement; it provides the mechanical resistance against which the muscles can act.

4. The microtubules and microfilaments in the effectors move by a system of ratchet-driven sliding motion.

OBJECTIVES

After studying this chapter and reflecting on it, you should be able to

1. Discuss the mechanism by which the hydrostatic skeleton functions, using as examples the leech and earthworm.

2. Distinguish between an exoskeleton and an endoskeleton, and list the advantages and disadvantages of each.

3. Show in a diagram the arrangement of the bones, muscles, tendons, and ligaments in a movable joint.

4. Differentiate among striated, smooth, and cardiac muscle with respect to cell structure, arrangement of cells in the muscle, innervation, type and duration of action, sensitivity to stimuli, resting length.

5. Contrast red muscle with white muscle.

6 Give a concise description of the physiology of muscle contraction at the nonmolecular level. Include in your answer the following terms: threshold, all-or-none, summation, tetanus, contraction, latent period, relaxation, fatigue.

7 Describe the internal fine structure of a skeletal muscle cell.

8 Explain the Huxley sliding-filament theory of skeletal muscle contraction.

9 Explain how the muscle cells get the ATP they need for contraction.

10 Describe the role played by calcium ions in muscle contraction.

11 Explain, using examples, the role of the actomyosin system in mediating nonmuscular movement.

12 Draw a diagram of a cross-section of a typical eucaryotic cilium or flagellum and discuss its probable mode of action.

SUMMARY

Although plants do move by differential growth or turgor changes, the most elaborate mechanisms for producing movement are found among animals. Movement in animals depends on either microtubules or microfilaments. The *effectors* are the parts of the organism that do things, that carry out the organism's response to a stimulus. The principal effectors of movement in higher animals are the muscle cells.

Movement in many invertebrate animals is produced by the alternating contraction of the longitudinal and circular muscles against the incompressible fluids in the body cavity. The fluids function as a *hydrostatic skeleton*. The hydrostatic skeleton of the earthworm is particularly efficient; the body cavity is segmented, and each segment has its own muscles. Consequently, the worm is capable of localized movement, and burrows very effectively.

Animals with hard jointed skeletons Both arthropods and vertebrates have evolved paired locomotory appendages, a hard jointed skeleton, and elaborate musculature. Arthropods have an *exoskeleton*, a hard body covering with all muscles and organs inside it. Besides providing support, the exoskeleton functions as protective armor and as a waxy barrier to prevent water loss. Periodic molting of the exoskeleton is necessary for growth.

Vertebrates have an *endoskeleton*, an internal framework with the muscles outside. It is composed of bone and/or cartilage. *Cartilage* is firm, but not as hard or brittle as bone. It is found wherever firmness combined with flexibility is needed. *Bone* has a hard, relatively rigid matrix and provides structural support and protection.

Vertebrate skeletons are divided into two components, the *axial* skeleton and the *appendicular* skeleton. Some bones are connected by immovable joints; others are held together at movable joints by *ligaments*. Skeletal muscles, attached to the bones by *tendons*, contract and bend the skeleton at these joints. The movable bones behave like a lever system with the fulcrum at the joint. The action of any specific muscle depends on the position of its *origins* (the ends attached to the stationary bone) and its *insertions* (the ends attached to the moving bone) and on the type of joint between them. The muscles operate in antagonistic and synergistic groups.

Vertebrates possess three different types of muscle: skeletal, smooth, and cardiac. (This classification does not hold for many invertebrates.) The abundant *skeletal muscle* is responsible for most voluntary movement. Each muscle fiber is long and cylindrical, contains many nuclei, and is crossed by light and dark bands called *striations*. The fibers are usually bound together into bundles, and the bundles into muscles. There are two types of skeletal muscle: *red muscle* (or slow-twitch muscle) and *white muscle* (or fast-twitch muscle). They differ in structure and function. Skeletal muscle is innervated by the somatic nervous system.

Smooth muscle forms the muscle layers in the walls of the viscera and the blood vessels. The spindle-shaped cells interlace to form sheets of tissue, innervated by the autonomic nervous system. Smooth muscle is primarily responsible for movements in response to internal changes, while skeletal muscle is concerned with making adjustments to the external environment. These differences in action are reflected in many differences in the physiological characteristics of these two muscle types.

Cardiac muscle (or heart muscle) shows some characteristics of skeletal muscle and some of smooth muscle. Its fibers are striated, but it is innervated by the autonomic nervous system, and it acts like smooth muscle.

The physiology of skeletal muscle activity Individual muscle fibers contract only if they receive a stimulus of threshold intensity, duration, and rate. In vertebrates, individual muscle fibers seem to exhibit the all-or-none property. When a single threshold stimulus is applied to a muscle, there is a brief *latent period*, which is followed by the *contraction period*. This is immediately followed by the *relaxation period*. These three periods comprise a single *simple twitch* of the muscle.

When frequent stimuli are applied to a muscle, the muscle does not have time to relax between contractions. The result is *summation*. If the stimuli are very frequent, the muscle may not relax at all between successive stimulations; the resulting strong sustained reaction is called *tetanus*. If the frequent stimulation continues, the muscle may fatigue and be unable to sustain the contraction.

The energy for muscle contraction comes from ATP. This energy comes from the complete oxidation of nutrients to CO_2 and H_2O. During violent muscular activity, the energy demands are greater than can be met by respiration, because oxygen cannot be gotten to the tissues fast enough. Consequently, lactic acid fermentation takes place, and the muscle incurs an *oxygen debt*. The panting that occurs even after the activity is over supplies the oxygen needed to reconvert the lactic acid into pyruvic acid and to metabolize it.

Analysis of muscle shows that its contractile components are the proteins *actin* and *myosin*. Within each *myofibril* the thick myosin filaments and thin actin filaments are interdigitated, with the myosin located exclusively in the *A bands* and the actin primarily in the *I bands* but extending some distance into the A bands. The contractile unit (between the *Z lines*) is called the *sarcomere*. According to the Huxley sliding-filament theory, cross bridges from the myosin filaments hook onto the actin at specialized receptor sites and bend, pulling the actin along the myosin. The necessary energy comes from the hydrolysis of ATP by the myosin cross bridges. As the filaments slide past each other, the width of the *H zone* and I band decreases.

The control of contraction The membrane of the resting muscle fiber is polarized, with the outside charged positively in relation to the inside. Stimulatory neurotransmitter substances released by the neuronal axon reduce this polarization, triggering an action potential that sweeps across the muscle fiber. When the action potential penetrates into the interior of the fiber via a *T tubule*, calcium ions are released by the terminal cisternae of the *sarcoplasmic reticulum*. The Ca^{++} ions stimulate contraction by binding to the protein *troponin* and causing a conformational change, which moves the *tropomyosin* and exposes the myosin-binding sites of actin. The filaments can now slide past each other by repeating the process of binding, movement, and release of the cross bridges. As long as Ca^{++} remains available, the muscle can continue to contract. When nervous stimulation ceases, the muscle relaxes because a calcium pump in the cisternal membrane moves the Ca^{++} back into the cisternae, and the myosin-binding sites are once again inhibited.

Effectors of nonmuscular movement Other types of eucaryotic cellular movement also depend on microfilaments, most of which also contain actin and a small amount of myosin. As in muscle, the complexing of actin and myosin activates the ATPase activity of the myosin, and contractions are stimulated by Ca^{++} ions. The same fundamental mechanism that is responsible for muscular movement appears to function in many other kinds of movement as well. The movements of microvilli, changes in cell shape, and cytoplasmic streaming all appear to be mediated by an actomyosin system.

Other movements in eucaryotic cells appear to be mediated by microtubules. Cilia and flagella contain a precise 9 + 2 arrangement of microtubules. Attached to the nine peripheral doublets are arms composed of the protein *dynein*, which has ATPase activity. The dynein arms may function as cross bridges between adjacent doublet microtubules, allowing them to slide over one another. Microtubules are also involved in the formation of the spindle in cell division, and in the movement of organelles within the neuronal axon.

QUESTIONS

Testing recall

The next two questions refer to the two bones in the drawing.

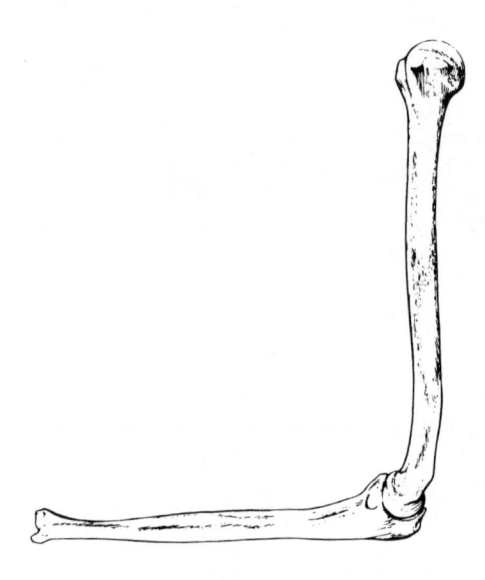

1. Draw in and label: ligaments, two antagonistic muscles (showing origins and insertions), tendons.
2. Do these bones form an endoskeleton, an exoskeleton, or a hydrostatic skeleton?

For items 3-12, indicate whether the statement is true of skeletal, smooth, or cardiac muscle. A statement may be true of more than one kind of muscle.

3. crossed by alternating light and dark bands
4. occurs in two types, red and white
5. each cell contains a single nucleus
6. fibers bound together into bundles by connective tissue
7. innervated by one kind of nerve fiber
8. innervated by the autonomic nervous system
9. can contract without nervous stimulation
10. capable of slow, sustained contraction
11. effects adjustment to the external environment
12. principal muscle type in insects

Use the kymograph record below to answer questions 13-16.

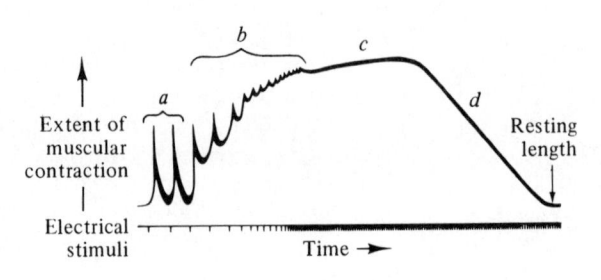

13. Referring to the diagram, state what is happening to the muscle in areas *a*, *b*, *c*, and *d*.
14. Why do several stimuli in rapid succession produce more contraction than a single stimulus of the same strength?
15. Are you using simple twitches or tetanic contractions when you write the answer to this question?
16. What causes fatigue?
17. Which of the following substances or processes are used by the cell to generate ATP for muscle contraction?

    a glycolysis   d creatine phosphate
    b fermentation   e myosin
    c cell respiration

18. Place the following seven events in the contraction of a muscle fiber in order.

    a Transmitter substance diffuses across synaptic cleft at neuromuscular junction.
    b Actin and myosin slide past each other.
    c Permeability of the muscle membrane is altered.
    d Impulse arrives at the junction between the motor neuron and the muscle fiber.
    e Sarcoplasmic reticulum releases stored calcium ions.
    f Action potential travels along membrane of the muscle cells and into the tubules.
    g Troponin changes shape and moves the tropomyosin.

For each of the following, indicate whether the movement is due mainly to: (1) microtubules or (2) microfilaments. Also specify the probable mechanism of movement: (a) actomyosin mediated, (b) dynein-tubulin mediated, or (c) mediated by some other mechanism.

19 cytoplasmic streaming

20 contraction of muscle

21 movement of mitochondria in the axon

22 beating of cilia

23 movement of ectoplasm through the endoplasm

24 movement of the microvilli

25 changes in cell shape

Testing knowledge and understanding

Choose the one best answer.

1 Vertebrate skeletal muscle is usually

   a under control of the sympathetic nervous system.
   b under control of the parasympathetic nervous system.
   c under control of the somatic (voluntary) nervous system.
   d dually innervated by sympathetic and parasympathetic neurons.
   e dually innervated by somatic and parasympathetic neurons.

2 If a single excised frog muscle is stimulated electrically and the responses are recorded on a kymograph drum, which one of the following *cannot* be demonstrated?

   a threshold   *d* fatigue
   b tetanus   *e* antagonism
   c summation

3 A sustained muscular contraction is

   a fatigue.   *d* oxygen debt.
   b a twitch.   *e* the latent period.
   c tetanus.

4 Muscular fatigue may occur with an accumulation of

   a ATP.   *d* myoglobin.
   b lactic acid.   *e* glucose.
   c oxyhemoglobin.

5 ATP in the muscle cell is regenerated directly from stores of

   a myosin.   *d* NADH.
   b ADP.   *e* pyruvic acid.
   c creatine phosphate.

6 The energy levels in the muscle are directly restored after muscle contraction by

   a cell respiration.
   b lactic acid formation.
   c ADP hydrolysis.
   d protein breakdown.
   e oxygen entry.

7 A muscle has built up an oxygen debt. When there is enough oxygen for aerobic respiration to resume, all of the following will occur except

   a lactic acid will be converted to pyruvic acid.
   b O_2 will be used up.
   c acetyl CoA will be converted to CO_2 and H_2O.
   d pyruvic acid will be converted to acetyl CoA.
   e an excess of $NADH_2$ will accumulate.

8 Once an impulse is transmitted across the neuromuscular junction by acetylcholine, continued transmission due to the same impulse is prevented by the breakdown of acetylcholine by

   a ATPase.
   b phosphatase.
   c acetylcholinesterase.
   d sodium-potassium pump.
   e acetylase.

9 The function of the T tubules in muscle contraction is to

   a carry the impulse into the myofibrils of the muscle cell.
   b release calcium ions.
   c release sodium ions.
   d split ATP.
   e take up calcium ions.

10 At the start of a muscle contraction, Ca^{++} is released from

   a tropomyosin.
   b troponin.
   c the motor neuron.
   d the sarcoplasmic reticulum.
   e myosin.

11 Which of the following chemicals are necessary for muscle contractions?

   a actin, myosin, calcium ions
   b actin, myosin, ADP
   c actin, ATP, myoglobin
   d myosin, calcium phosphate, ATP
   e myosin, calcium ions, myoglobin

12 Which of the following statements is not true?

   a In skeletal muscles, there is a regular alternation of light-colored I bands and dark-colored A bands.
   b It is now thought that the A bands correspond to the lengths of thick myosin filaments.
   c When a muscle contracts, the A bands become much shorter whereas the I bands remain roughly the same length.
   d It is thought that cross bridges from the myosin filaments, acting like ratchets, provide the pull that slides the filaments together during contraction.
   e The direct stimulant for contraction is thought to be release of intracellular calcium from cisternae of the sarcoplasmic reticulum.

13 When a skeletal muscle contracts, all of the following occur *except*

   a the H zone almost disappears.
   b the Z lines move closer together.
   c the I bands diminish markedly in width.
   d ATP is hydrolyzed to ADP + P_i.
   e the A bands diminish markedly in width.

For further thought

1 The earthworm has exploited the hydrostatic skeleton to its fullest. How might the earthworm's movements differ if the worm were not segmented?

2 Why does a person pant and sweat after vigorous exercise?

3 Explain what happens when rigor mortis sets in after death.

4 In a chicken or turkey, the breast is white meat (white muscle) whereas the leg is dark meat (red muscle). In many migratory birds the situation is reversed: the breast meat is red muscle, the legs are white. Explain this difference.

Chapter 12

ANIMAL BEHAVIOR

KEY CONCEPTS

1. Inheritance and learning are both fundamental in determining the behavior of higher animals; their contributions are inextricably intertwined in most behavior patterns.

2. All forms of learning and performances of behavior patterns depend on adequate motivation in the animal.

3. Sound is an effective means of communication among many animals; by varying the pitch, frequency, and volume an amazing amount of information can be conveyed.

4. The olfactory sense is immensely important in the lives of many animals and constitutes a basis for effective communication.

5. Visual displays are very important in reproductive and agonistic behaviors; they communicate the current balance between an animal's attack and escape motivation.

6. All living things, whether individual cells or whole multicellular organisms, appear to have an internal sense of time—a biological clock.

7. Effective feeding behavior can be analyzed in terms of energy: the energy gained from the food must be greater than that invested in obtaining the food.

8. The degree of sociality among animals is correlated with their ability to communicate with one another.

9. Behavior is adaptive; natural selection brings about an increase in well-adapted and a decrease in poorly adapted behavior patterns.

OBJECTIVES

After studying this chapter and reflecting on it, you should be able to

1. Define behavior and indicate why the analysis of behavior poses problems for the ethologist.

2. State Occam's razor and Morgan's canon and explain why these are important principles in interpreting behavior. Tell why these principles should not be applied too rigidly to complex behavioral patterns in higher animals.

3. Differentiate among taxis, kinesis, reflex, and instinct.

4. Summarize the current thinking on the relationship between inheritance and learning in animal behavior.

5. State the five major factors that complicate the study of learning.

6. Define and give an example of each of the five different types of learning.

7. Discuss motivation and drives and explain how behavior functions to maintain homeostasis within the individual organism.

8. Describe some of the characteristics of sign stimuli and their role in animal behavior.

9. Discuss the relative advantages and disadvantages of visual, auditory, and chemical means of communication.

10. Define displays, and discuss the importance of reproductive and agonistic displays.

11. Describe the honeybee's round dance and waggle dance, and give the information each conveys.

12. Explain the terms biological clock and circadian rhythm, and explain how the environment can reset the clock.

13. Discuss the advantages of each of the following mechanisms of orientation, and give examples of animals that utilize each type: visual landmarks, echolocation, electric orientation, and olfactory cues.

14. Discuss how birds orient themselves on long flights, and explain the factors that appear to be involved in the pigeon's navigational system.

15. Explain how the various components of feeding behavior can be analyzed on the basis of energetic considerations.

16. Describe and give advantages of the social organization seen in honeybee hives and ant nests.

17. Give three differences between insect societies and vertebrate societies.

18. Define a hierarchy in vertebrate social organization, give the advantages of this system, and distinguish between peck right and peck dominance.

19. Describe the different vertebrate mating systems and the role of parental investment in determining the evolution of these systems.

20. Describe the methods the biologist uses to study the evolution of behavior patterns.

21. Differentiate among displacement activity, redirected activity, and intention movements, and explain how these provide raw material for displays.

22. Cite an experiment showing that the genetic basis of many behavior patterns can be studied with the same sort of analysis as the genetic basis of anatomical and physiological traits.

SUMMARY

The behavior of an animal—what it does and how it does it—is the product of the functions and interactions of its various control and effector mechanisms. Behavior is a synthetic subject, depending for its own advances on advances in many other disciplines.

Analysis of behavior The analysis of behavior poses a problem because it is difficult to avoid anthropomorphic interpretations. Two principles are important in this connection: *Occam's razor*, which states that explanations should be no more complicated than necessary, and *Morgan's canon*, which says that we should interpret animal behavior in terms of the simplest neural mechanism that can explain the observed behavior. Morgan's canon should not be applied too rigidly to complex behavior patterns in higher animals, however. Taxes, kineses, and reflexes are types of simple behavioral patterns. A *taxis* is a continuously oriented response to a stimulus; a *kinesis* involves movements not oriented to the stimulus; and a *reflex* is a simple, automatic response to a stimulus. Higher animals show more complicated behavioral patterns. Patterns that are rather rigid and stereotyped are known as *instincts*.

Inheritance and learning are both fundamental in determining the behavior of higher animals; their contributions are inextricably intertwined in most behavior patterns. Inheritance sets the limits within which a behavior can be modified, and learning determines, within those limits, the precise character of the behavior. In general, the inherited limits within which behavior patterns can be modified by learning are much narrower in the invertebrates than in the vertebrates, and narrower in the lower vertebrates than in the mammals. In any animal the limits for different behavior patterns are different; each animal has some traits that are rather rigidly determined by inheritance, and others that are capable of much modification.

Several factors complicate the study of learning.

1 Often it is difficult to determine whether improvement is the result of learning, maturation, or physiological change.

2 An animal may learn something in one context and not in another.

3 An animal can learn certain behavior patterns only during a rather limited *sensitive phase* in its life.

4 Often there is a delay between the *latent learning* and the performance of the behavior pattern.

5 Comparisons between the learning capabilities of different species may be ill-founded and are often misleading.

Learning refers to relatively enduring changes in behavior due to experience rather than maturation; several types of learning can be distinguished. *Habituation* is a simple type of learning in which the organism shows a gradual decline in response to repeated insignificant stimuli that bring no *reinforcement*. *Conditioning* is the association of a response with a stimulus with which it was not previously associated. In *trial and error* learning the animal learns to associate a certain activity with a reward or punishment. In *imprinting* an animal learns to make a strong association with another organism (or object); imprinting occurs only during a short sensitive phase. Imprinting appears to be important in parental recognition and in proper species recognition and interaction. *Insight* learning is the organism's ability to respond correctly the first time it encounters a novel situation; it is most prevalent in higher primates. Insight must be distinguished from *generalization*, which is the ability of an animal conditioned to respond to one stimulus to respond in the same way to other similar stimuli.

Motivation can be defined as the internal state of an animal that is the immediate cause of the behavior. Both learning and performance depend on adequate motivation. Motivation is often thought of in terms of drives (e.g. hunger, thirst, sex). Many factors such as health, degree of motivation, previous experience, and hormones determine a drive. Motivated behavior is goal-directed; the behavior tends to lead to the fulfillment of biological needs. Goal-directed behavior has three components: searching or *appetitive* behavior, the *consummatory* act, and the *acquiescent* period. Such behavior contributes to the maintenance of homeostasis in the individual organism.

An animal responds to a limited number of the thousands of stimuli its receptors are detecting. Those stimuli that are particularly effective in triggering behavior are called *sign stimuli*. The animal must possess neural mechanisms that are selectively sensitive to the sign stimuli; these *releasing mechanisms* initiate the behavior when activated by the appropriate sign stimuli. Often it is possible to design *supernormal stimuli* even more effective than the natural ones. The intensity of the sign stimulus necessary to trigger a response is inversely proportional to the animal's motivation to perform the behavior. Also, some responses to sign stimuli occur only during a sensitive phase in the animal's life.

Animal communication Many animals can communicate information by sound. In many insects, sound functions as a mating call and a species recognition signal. Frog calls serve a similar function. The call of a male frog is an effective stimulus for the female only if the female's reproductive drive is high. Many frogs exhibit regional dialects in their songs.

Bird song functions as a species recognition signal, as an individual recognition signal, as a display that attracts females to the male and helps synchronize their reproductive drives, and as a display to defend territory.

Chemicals provide an effective means of communication among many organisms. Many animals secrete substances (*pheromones*) that influence the behavior of other organisms. Most pheromones can be classified into two groups: *releasers*, which trigger a rather immediate and reversible behavioral change in the recipient (e.g. sex attractants in insects); and *primers*, which initiate more profound long-term physiological changes in the recipient. Primers may not produce any immediate behavioral change.

Visual displays are frequently important in reproductive behavior. (A display is a behavior that has evolved specifically as a signal.) They function both in bringing together and synchronizing the two sexes in the mating act, and in avoiding mating errors.

Agonistic encounters between individuals of the same species may often be resolved by vocal and/or visual displays without physical combat. Through displays the combatants convey their attack motivation; the individual showing the higher attack motivation usually wins. In visual displays the loser often responds with *appeasement displays*; these tend to inhibit further attacks by the antagonist. Displays communicate the current balance between an individual's attack and escape motivations.

Scout honeybees are able to communicate to workers in the hive the quality of a food source as well as its distance and direction from the hive. When food is close to the hive the bees dance a round dance, which means nearby food. When food is farther away a waggle dance is performed; its length and orientation convey information on the distance of the food source and its direction relative to the sun's position. The symbolic communication of bees and chimpanzees suggests that symbolic communication is not exclusive to human beings.

Biological clocks All living things, whether individual cells or whole multicellular organisms, appear to have an internal sense of time—a biological clock. Many organisms show *circadian rhythms*, activity rhythms that vary within a period of approximately 24 hours. Although the basic period of the clock is innate, it is constantly being reset by the environmental cycle. When an organism is exposed to new conditions that shift the setting of its clock, the clocks of the various cells and organs do not shift together; thus the different organs may be thrown out of phase with one another.

Orientation behavior Many animals find their way from place to place by using *visual landmarks*. Other animals use *echolocation* to orient. For example, bats emit pulses of high-frequency sounds and adjust their flight by detecting the echoes from obstacles that return to their ears. *Electric orientation* is used by six groups of fishes; they produce weak electric fields that help them find their way about and locate prey. Many animals use *olfactory* cues for orientation. Most mammals receive olfactory information from hundreds of sources every minute; they recognize objects most clearly by smell.

Migrating birds and homing pigeons have remarkable navigational abilities. Birds (and many other animals) can tell compass directions by observing the sun and stars. Recently it has been shown that pigeons can use magnetic cues in addition to sun cues for orientation. Magnetic responses have now been found in birds, insects, a variety of invertebrates, and some fishes and amphibians. Other cues such as olfactory cues, meteorological parameters, and special sensory capabilities (e.g. barometric pressure detection, infrasound detection) may play some role in avian orientation.

Feeding behavior The behavior patterns associated with feeding vary with the animal's mode of life and type of food. Most animals can be characterized as either *specialist* or *generalist* feeders. The behavior can often be analyzed in terms of energetic considerations. The energy gained from eating an organism must be greater than the energy required to find, capture, and eat it. Therefore, efficient feeding behavior involves deciding where to search, how long to search in the selected area, what prey to search for (formation of *search images*), and whether to pursue a prey individual.

Social behavior Many animal species show some degree of intraspecific cooperation. The cooperation varies from relatively simple and transient forms to highly evolved and long-lasting societies in which almost every aspect of each individual's life depends on the activities of others. Effective communication between individuals is a prerequisite for such cooperation.

The spatial arrangement of individuals influences the species' social behavior. In many species each individual is surrounded by a small volume of space that is his own—the *individual distance*. Moderately social animals maintain a relatively inviolate individual distance; it tends to be less rigid in highly social animals. A *territory* is an area an animal or group of animals actively defends from members of the same species. By spacing individuals, territoriality minimizes competition and reduces social conflict. The *home range* is the largest spacing unit, the total area in which an animal (or group of animals) normally travels.

In the social insects the unit of organization is invariably the family, which typically consists of a single reproductive female, the queen, and her daughters (sometimes her sons). Different species form societies of varying complexity. The honeybee and ant societies are highly developed. Division of labor is based on biologically determined castes. The queen's role is reproductive; the sterile workers build the hive or nest, nurse the young, gather the food, and care for the queen.

A vertebrate social group typically consists of a leader together with his mates and offspring. Such a grouping benefits both from reduced aggression and from the greater tendency for cooperativity and altruism to evolve among individuals with shared genes. There is no biologically determined caste system, and all adults are potential reproducers.

An important aspect of vertebrate social organization is the *hierarchy*, a series of dominance-subordination relationships that order the group. Such factors as age, strength, size, health, seniority, and location of the first encounter help determine the animal's hierarchical position. In hens a rigid *peck right* system is established; the dominant individual can peck subordinate individuals without being pecked back. In other animals the hierarchy may be more fluid, as in the *peck dominance* system in pigeons. Social hierarchies tend to give order and stability to the group.

Several different mating systems are found in vertebrates. *Monogamy*, in which one male mates with one female, is common in birds but rare in mammals. There are two types of *polygamy: polyandry*, in which one female mates with many males; and the more common *polygyny*, in which one male mates with many females. Many species form no pair bonds; their mating is often *promiscuous*.

One of the principal determining factors in the evolution of mating systems is *parental investment*—the cost to the parent of behavior enhancing the likelihood that the offspring will survive and reproduce. In songbirds monogamy is adaptive because the young are born helpless and it may require a large parental investment from both parents to raise them. In mammals the male and female have very unequal parental investments and polygyny may be an effective reproductive strategy.

The evolution of behavior Behavior is adaptive; natural selection brings about an increase in well-adapted and a decrease in poorly adapted behavior patterns in the population. Behavior patterns evolve and can be studied in terms of the selection pressures that produce them. By comparing the behavior patterns of a whole group of related species, the evolutionary derivation of the patterns can be clarified. The evidence from behavioral analysis is also useful.

Detailed analysis of behavior patterns can help identify the actions from which the behavior is derived. *Displacement activity* (irrelevant action), *redirected activity* (an appropriate action directed at the wrong object), and *intention movements* (incomplete initial stages of other actions) provide raw materials for displays. Natural selection tends to exaggerate and ritualize these actions, increasing their information content and thus giving them a display function.

Studies of the genetics of behavior indicate that, despite the complicating modification of behavior by learning, the underlying genetic basis of many behavior patterns can be analyzed with some of the traditional procedures of geneticists.

QUESTIONS

Testing recall

In the statements below, underline the correct word or words.

1 The statement that "if several explanations are all compatible with the evidence at hand, the simplest one should be considered most probable" is called (Occam's razor, Morgan's canon).

2 An animal's simple movement toward or away from a stimulus is called a (tropism, taxis, kinesis, reflex).

3 The behavior of higher animals depends on (inheritance, learning, motivation).

4 The inherited limits within which behavior can be modified are much (wider, narrower) in the invertebrates than in the vertebrates.

5 If you catch a wild snake and handle it, after a while it will be calm when you hold it. This is an example of (habituation, conditioning, trial-and-error learning, imprinting).

6 A form of learning that ordinarily occurs quickly but during only a limited phase in the animal's life, happens quickly, and is difficult to unlearn is called (conditioning, insight, habituation, imprinting).

7 The area of the brain that acts as the center for behavioral drives in higher vertebrates is the (cerebral cortex, hypothalamus, cerebellum, midbrain).

8 Bird songs may function (to indicate happiness, in territorial defense, in species recognition, as a display).

9 Chemical substances released by animals that produce long-term physiological changes in other animals of the same species are called (primer pheromones, releaser pheromones, hormones).

10 A dog approaches another with head up, hackles raised, and teeth bared. This is a(n) (greeting, appeasement, threat, mating) display.

11 Circadian rhythms have a period of about a (day, week, month, year).

12 A dolphin swimming in murky water would find it difficult to orient if you covered its (eyes, ears, electric organ).

13 Homing pigeons can use (the sun compass, the star compass, magnetic cues) to orient home.

14 Animals that feed on a wide variety of foods are called (specialist, generalist) feeders.

15 When an animal concentrates on finding one type of prey it has formed a (search image, feeding specificity).

16 The total area in which an animal normally travels is known as its (individual distance, territory, home range).

17 In vertebrates, social hierarchies (give order and stability, decrease aggression).

18 The mating system in which one female mates with many males is called (monogamy, polygyny, polyandry).

19 Behavior patterns (do, do not) evolve.

20 When an animal performs an irrelevant action in a conflicting situation it is a(n) (displacement activity, redirected activity, intention movement).

Testing knowledge and understanding

Choose the one best answer.

1 Behavior can be modified by
 a the nature of the sign stimulus.
 b the season of the year and the time of day.
 c the level of circulating hormones.
 d previous experience.
 e all the above.

2 In an experiment, a single frog tadpole was raised in an isolated, soundproof box. As soon as the tadpole transformed into a frog, several adult males of the same species were put into the box for several days; they called continuously during this period and were then removed. Except for these males, the original frog was never exposed to another frog. At maturity, it produced a call typical of its species. From the

information provided here, what can you say about the basis of this call?

a The call is entirely learned.
b The call is entirely innate.
c The call is partly learned, partly innate.
d The call is neither learned nor innate.

3 When a blowfly maggot is exposed to light it moves directly away from the light source. This is a(n)

a reflex.　　　　　*d* instinct.
b taxis.　　　　　*e* learned behavior.
c kinesis.

Questions 4–8 describe experiments or observations that can be classified as one of the forms of learning listed below. Use this set of answers for all questions; a given answer may be used more than once or not at all.

a insight　　　　　*d* trial and error
b habituation　　　*e* imprinting
c conditioning

4 A duckling will follow any large moving object presented to it right after hatching.

5 Adult White-crowned Sparrow males "defend" their territorial boundaries by vigorously singing their characteristic songs. If a tape-recorded sparrow song is played nearby, the male will first search frantically for the "other bird" and then sing his song at the speaker. When the tape is played repeatedly, the sparrow eventually gives up singing and resumes his normal activities.

6 A chimpanzee is put in a cage with several boxes scattered across the floor and a clump of bananas hung from the top of the cage, out of reach. The monkey piles the boxes on top of each other, climbs up, and eats the bananas.

7 Flatworms respond to an electric shock by contracting their body muscles. If the worms are exposed to 50 or 100 electric shocks and at the same time to a beam of light, the worms will learn to contract immediately upon presentation of the light beam even if no shock is present.

8 A hungry brown rat was placed in a closed metal box with a food slot and a food-releasing lever. The rat was allowed to poke around randomly until it accidentally tripped the lever and released a food pellet. The rat soon learned to trip the lever whenever it wanted food.

9 Which one of the following types of learning is probably least widespread in the animal kingdom?

a trial-and-error learning
b habituation
c conditioning
d insight learning

10 In human beings, certain perceptions of tactile stimuli release specific defensive responses; for example, creeping things on the back of the hand release a shaking movement of the hand. This response is probably due to

a innate or instinctive behavior.
b imprinting.
c habituation.
d conditioning.
e insight learning.

11 A bird is hatched and raised indoors in complete isolation from other members of its own species. It becomes imprinted on the student who raised it. When the bird is released to the wild as an adult, which one of the following behaviors would you *not* expect it to have difficulty performing correctly?

a singing its species-typical song
b using star patterns in migratory orientation
c finding a mate
d flying

12 Which one of the following statements about imprinting is *true*?

a Responsiveness to the releasing object is lost after the sensitive period is over.
b It occurs only in mammals.
c It is typically reversible, in both lab and natural situations.
d It occurs early in the animal's life.

13 When ready to reproduce, the adult male stickleback fish develops a bright red belly, which he displays aggressively to other males when staking out his nesting territory. Cardboard models of the fish placed in a nesting male's territory also elicit an aggressive display if the lower half of the model is painted red. The red belly is an example of a(n)

a reflex.　　　　　*d* learned pattern.
b sign stimulus.　　*e* taxis.
c instinct.

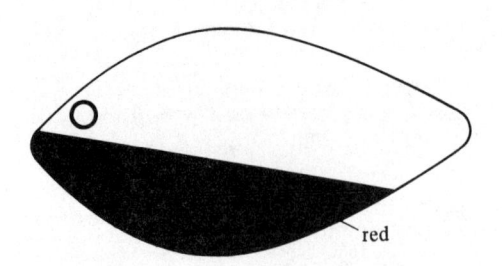

14 The courting male stickleback fish responded most vigorously to an extra-large model that looked like the figure. This preferred but unnatural stimulus is sometimes referred to as a

a releaser.
b supernormal stimulus.
c kinesis.
d displacement activity.
e misdirected effort.

15 Which of the following methods of communication would be most adaptive as a means of intraspecific communication for an organism having these characteristics: it is nocturnal, terrestrial, lives individually in deep holes, and has a large number of potential predators which could extract it from its burrow. Other members of this species live in nearby holes.

a olfactory cues
b thermal cues
c visual cues
d auditory cues
e electrical cues

16 Which one of the following is *not* true of circadian rhythms?

a The rhythmic activity pattern often persists for a long period even if the environment is kept constant.
b The rhythmic pattern can be reset or entrained to an altered pattern of light and dark cycles.
c When the "clock" is reset to coincide with a new environment all the physiological functions which have a circadian rhythm readjust simultaneously.
d The rhythm appears to be innate.
e The clock is not set precisely to a 24-hour day; it must be reset each day to keep it in syncrony with true time.

17 On which of the following flights would you expect passengers to experience the most severe cases of jet lag?

a New York to San Francisco
b Cleveland to Miami
c Chicago to San Francisco
d San Francisco to New Orleans
e San Francisco to Chicago

Questions 18-20 refer to the diagram showing locations of feeding sites (A through C) and the hive, as well as the position of the sun. Assume that the sun remains stationary.

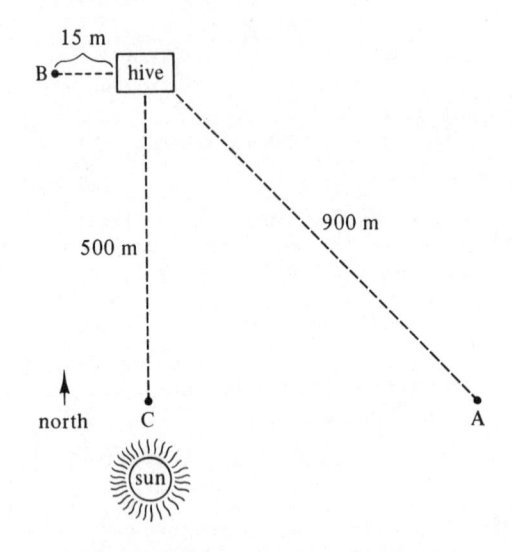

18 If a scout bee returned to the hive from a successful flight to site A and performed its dance on a *vertical* comb, what would you expect its dance to look like?

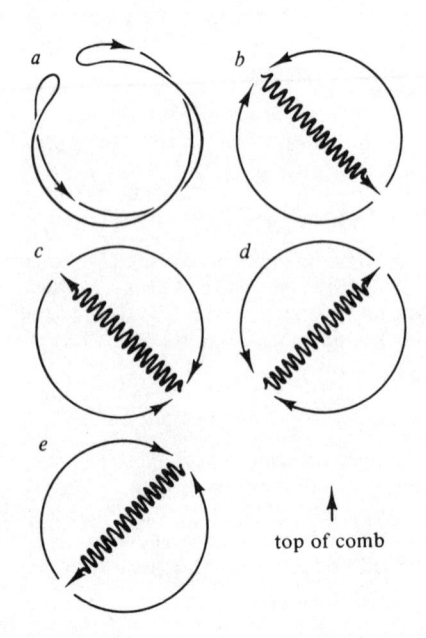

19 If another bee returned to the same hive from site B and danced on a *vertical* comb, what would you expect its dance to look like?

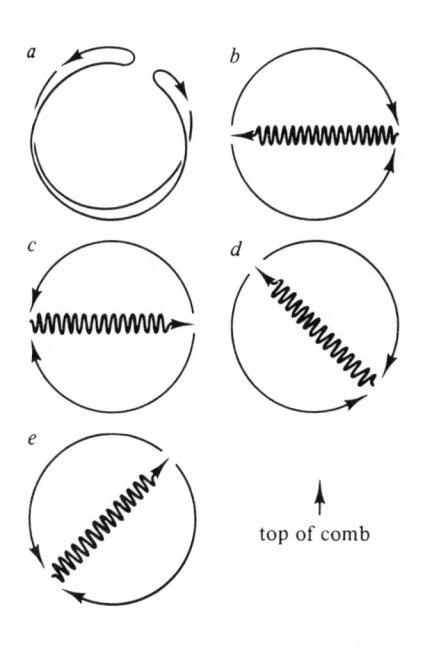

20 If another bee returned to the same hive from site C and performed its dance on a *horizontal* platform in front of the hive, what would you expect the dance to look like?

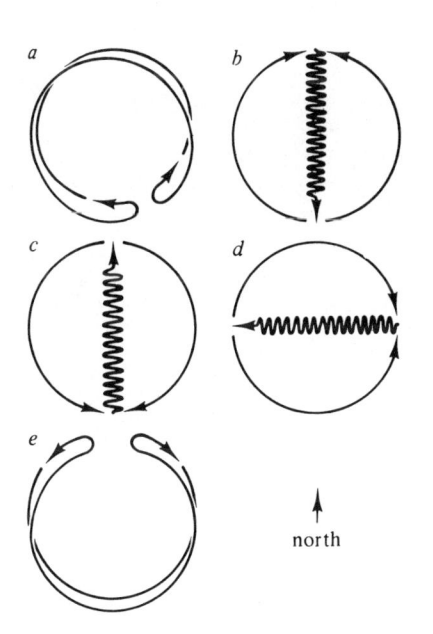

21 In the waggle dance of a honeybee on a vertical comb in a darkened hive, the distance of the food source is communicated by the

  a angle between the waggle run and the vertical.
  b length of the waggle run.
  c number of runs per unit time.
  d angle between the waggle run and the hive entrance.
  e number of waggles per run.

Questions 22–24 refer to the diagram. Several birds of the same species are kept in an outdoor cage and released to determine in which direction they would migrate after various experimental manipulations as described in the questions. Use the following answers for all three questions; an answer may be used more than once or not at all.

  a north   d west
  b south   e northeast
  c east

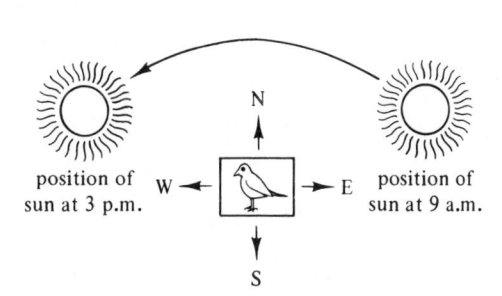

22 In a test conducted at 9 a.m. it is found that the migratory direction is *south*. If a bird is then kept outdoors so that it can see the sun throughout the day and is retested six hours later, at 3 p.m., in which direction would you expect it to migrate?

23 The same bird is tested again but immediately after its 9 a.m. test is put in a light-tight black box for a day; during this time it cannot see the sun. After such treatment, the bird is tested at 3 p.m. under the sun. In which direction would you expect it to migrate?

24 During the experiment, several birds are born in the cage; these have never flown anywhere before. If such a bird is exposed to the normal daytime sky, in what direction would you expect it to go when released? Assume that these birds are native to the area in which they are caged and released.

25 Which statement is *not true* of the navigation of homing pigeons?

   a Homing pigeons are not the only animals that can return home when displaced.
   b Pigeons with the phase of their internal clock shifted by six hours orient homewards on sunny days.
   c Pigeons with their internal clock shifted by six hours orient homewards on overcast days.
   d Magnets can disorient pigeons on overcast days.

26 Wolves urinate to mark the edges of their territories. This is an example of

   a a reflex.
   b using pheromones.
   c marking individual distance.
   d mating behavior.
   e marking home range.

27 A fighting cock is faced with a male turkey. Instead of fighting each other or fleeing, both begin to peck vigorously at the ground. This behavior may be best interpreted as an example of

   a kinesis.
   b displacement activity.
   c redirection activity.
   d taxis.
   e tropism.

28 Which one of the following statements is *not* true?

   a Courtship displays help synchronize the reproductive physiology of male and female.
   b Courtship displays by males help them attract females.
   c Courtship displays are often important in species recognition in areas where two or more closely related species are found together.
   d Agonistic displays often enable animals to resolve hostile encounters without physical combat.
   e Agonistic displays are seldom effective in resolving disputes over territorial boundaries.

29 Two dogs have had an agonistic encounter and one of them is beginning to show appeasement displays. Which one of the following actions would you *not* expect to be a part of such appeasement displays?

   a The tail is tucked between the legs.
   b The hair is sleeked.
   c The stance of the body is lowered.
   d The head is turned away from the opponent.
   e The ears are held erect.

30 In a particular animal species the young are precocial; they can help care for themselves almost immediately. The care and feeding of the young is done by the female parent, although the male does guard the female and young. The most likely mating system for this species is

   a monogamy.
   b polyandry.
   c polygyny.
   d promiscuity

31 Which one of the following is true of *both* insect and vertebrate societies?

   a The unit of organization is usually the family.
   b Division of labor is based on biologically determined castes.
   c All individuals in the society are capable of reproduction.
   d Individual recognition between members is crucial to the organization.
   e The organization is based on a rigid hierarchical system.

For further thought

1 A bird in Ithaca, New York is trained to fly south and uses the sun as a reference point; normal sunrise is at 6 a.m. The bird then spends several weeks in a lab chamber where artificial lights come on six hours *later* than sunrise (although the number of "daylight" hours remains the same). In which compass direction would you expect this phase-shifted bird to fly if released at noon in Ithaca? Assume that the sun is in view.

2 Another bird of a different species normally migrates south from Ithaca at night, using star patterns to maintain its bearing. Normal sunrise is at 6 a.m. This bird spends several weeks in a lab chamber where artificial lights come on six hours *earlier* than sunrise (although the number of **"daylight" hours remains the same**). In which compass direction would you now expect this phase-shifted bird to fly if released at midnight in Ithaca? Assume that the stars are in view.

3 Kramer showed the importance of the sun as a reference point for bird orientation with a simple experiment in which he blocked off the view of the sun and used mirrors to make it appear in other positions relative to a bird in an experimental arena. If the position of the true sun with respect to the bird's migratory direction (indicated by the heavy arrow) is shown on the left, in which direction (*a* through *e*) would you expect the bird to attempt to fly if the sun's position is artificially altered as depicted on the right?

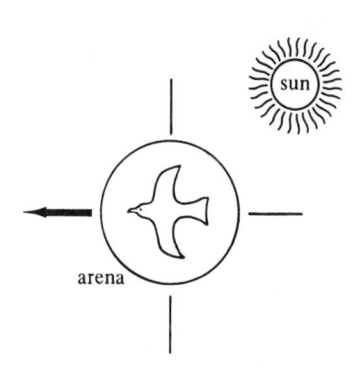

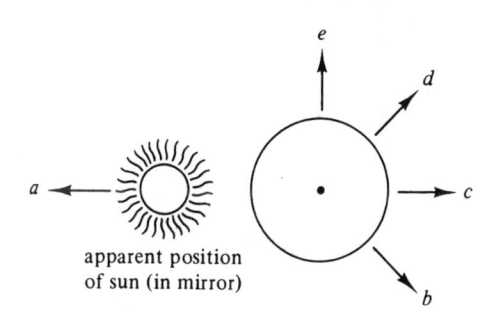

4 Over the years there has been a good deal of controversy concerning innate versus learned behaviors. Is it possible to distinguish between them? Explain.

5 When a young kitten plays it frequently pounces on moving objects. As an adult it will use this same method to catch mice. Has the cat "learned" to catch mice or is catching mice a result of maturation? Design an experiment that would help you answer this question.

6 Scientific investigation has shown that human females when housed together tend to develop synchronous menstrual cycles. Discuss mechanisms that might account for this.

7 Discuss current ideas about the "raw materials" from which displays may evolve.

8 Discuss the importance of displays in animal courtship and/or agonistic encounters, giving examples of actual displays.

9 Explain why differences in courtship displays could play an important role in preventing hybridization between two closely related species living in the same area.

Chapter 13

CELLULAR REPRODUCTION

KEY CONCEPTS

1 When a cell divides by mitosis, it must make a complete copy of the genetic information in its nucleus and then, as it divides, give one complete set to each daughter cell.

2 Mitosis produces new cells with exactly the same chromosomal endowment as the parent.

3 Meiosis reduces the number of chromosomes by half so that when the egg and sperm unite in fertilization, the normal diploid number is restored.

4 Sexual reproduction increases variation in the population by making possible genetic recombination.

OBJECTIVES

After studying this chapter and reflecting on it, you should be able to

1 Cite experimental evidence showing the essential role of the nucleus in the life of the cell and in the transmission of hereditary information.

2 Define binary fission and give examples of organisms that reproduce by this process.

3 Distinguish between mitosis and cytokinesis.

4 Diagram the cell cycle, including the different stages of interphase and of mitosis.

5 Discuss the events that occur during each stage of mitosis. Differentiate between mitosis in the plant cell and in the animal cell.

6 Distinguish between haploid and diploid cells.

7 Compare and contrast meiosis with mitosis, and give the importance of each process in the life of the organism.

8 Explain the function of the second meiotic division.

9 Explain why spermatogenesis produces four sperm cells whereas oogenesis produces only one egg.

10 Indicate how the products of meiosis differ in plants and animals.

11 Diagram the three types of life cycles and name an organism that exemplifies each life cycle.

12 Discuss reasons why natural selection has favored the more complicated process of sexual reproduction over the simpler one of asexual reproduction in a wide variety of organisms.

SUMMARY

The nucleus is the control center of the cell; it contains the genetic information that determines the organism's characteristics and directs its activities. The nucleus is essential for the life of the cell; normal function depends on it.

Procaryotic cells divide by *binary fission*; the single circular chromosome replicates and a ring of new plasma membrane and wall material grows inward, separating the replicates.

Mitotic cell division Cell division in eucaryotic cells involves two processes: the division of the nucleus (*mitosis*) and the division of the cytoplasm (*cytokinesis*). Nuclear division entails, first, precise duplication of the genetic material and, second, distribution of a complete set of the material to each daughter cell.

Cells that are going to divide pass through a series of stages known as the *cell cycle*. After a cell has completed the division process, there is a gap in time, the G_1 stage, before replication of the genetic material begins. Next comes the S stage, during which new DNA is synthesized. Another time gap, the G_2, separates the end of replication from the onset of mitosis proper.

Some animal cells become arrested in the G_1 stage and a few in G_2; these cells normally will not divide. The arrests appear to be due to failure to produce some control chemical. Production of these control substances depends on stimulation from certain specific peptide *growth factors* from blood serum.

Mitosis can be inhibited by chemicals called *chalones*; these peptide or glycoprotein substances act on specific target tissues. The balance between the chalones and growth factors determines whether the cell will divide.

During the G_1, S, and G_2 stages, the cell is nondividing; these three stages together constitute the *interphase* state. During interphase the nucleus is visible, and one or more nucleoli are prominent. The chromosomes are long and thin and cannot be seen as distinct structures. After the cell has passed through G_1, S, and G_2 it enters mitosis. Mitosis is customarily divided into four stages:

1 *Prophase*. As the two centrioles move to the opposite sides of the nucleus, the chromosomes condense into visible threads. Each chromosome consists of two *chromatids* held together by their *centromeres*. The *spindle* and *astral* microtubules appear near each centriole. The nuclear membrane and nucleoli disappear. Plant cells, unlike animal cells, lack centrioles and astral microtubules.

2 *Metaphase*. The centromeres of each double-stranded chromosome attach to a spindle microtubule along the equator of the cell. Metaphase ends when the centromeres of each double-stranded chromosome uncouple.

3 *Anaphase*. The two new single-stranded chromosomes separate, one moving toward each pole. Although they are moved by the chromosome-to-pole microtubules, the mechanism of this movement is unknown. Cytokinesis often begins here.

4 *Telophase*. The chromosomes reach the poles, the nuclear membrane and nucleoli are reformed, cytokinesis is completed, and the chromosomes become longer and thinner.

Cytokinesis frequently accompanies mitosis. In animal cells, cytokinesis begins with the formation of a *cleavage furrow* running around the cell; the furrow deepens until it divides the cell in two. In plants, a *cell plate* forms in the center of the cytoplasm and enlarges until it cuts the cell in two.

Mitosis produces two new cells with exactly the same chromosomal endowment as the parent cell.

Meiotic cell division *Meiosis* is a special process of cell division that reduces the number of chromosomes by half so that when the egg and sperm unite in fertilization the normal number is restored. During the reduction division of meiosis, the chromosomal pairs are partitioned so that each gamete contains one of each type of chromosome. (It is *haploid*.) When the two gametes unite in fertilization, the resulting zygote is *diploid*, having received one chromosomal type from each parent.

Meiosis involves two successive divisional sequences, which produce four new haploid cells. The first division (meiosis I) is the reduction division; the second (meiosis II) separates the chromatids. Both meiosis I and II can be divided into four stages. In meiosis I these are

1 *Prophase I.* The events are similar to the mitotic prophase, except that in meiosis the members of each pair of chromosomes (homologous chromosomes) come together in a process known as *synapsis*. Sometimes chromatids of homologous chromosomes exchange parts in a process called *crossing-over*.

2 *Metaphase I.* The two chromosomes of each homologous pair attach to the same spindle microtubule along the equator of the cell. The centromeres do not uncouple.

3 *Anaphase I.* The two double-stranded chromosomes of each synaptic pair move to opposite poles.

4 *Telophase I.* Two new nuclei are formed, each with half the chromosomes present in the parental nucleus. The nuclei are not identical.

A short period called *interkinesis* follows telophase I. No replication of genetic material occurs during this stage. The second division sequence of meiosis is essentially mitotic; each double-stranded chromosome attaches to a *separate* microtubule, the centromeres uncouple, and the new single-stranded chromosomes move to the poles. Four new haploid cells containing single-stranded chromosomes are produced. Meiosis II is necessary to complete the segregation of recombinant genetic entities that result from crossing-over.

Higher animals are diploid. During reproduction, meiosis produces haploid gametes, which unite to produce the diploid zygote. *Gametes* are haploid cells specialized for sexual reproduction. In male animals, *spermatogenesis* produces four functional sperm cells from each diploid cell. In females, *oogenesis* produces only one mature ovum; the polar bodies are nonfunctional.

Meiosis in plants produces haploid cells called *spores* (stage 1), which often divide mitotically to form haploid multicellular plants (stage 2). Eventually these haploid plants produce gametes (stage 3) by mitosis. Two of the gametes unite to form the diploid zygote (stage 4), which develops into a diploid multicellular plant (stage 5), and the cycle is complete. Most multicellular plants have all five stages in their life cycles, but the relative importance of the stages varies greatly. In general, the haploid stages are dominant in the primitive plants and the diploid in the more advanced plants.

The adaptive significance of sexual reproduction Mitotic asexual reproduction produces new cells genetically identical to the parent cell, whereas sexual reproduction increases variation in the population by making possible genetic recombinations. Each diploid cell undergoing meiosis can produce 2^n different chromosomal combinations, where n is the haploid number. The number of possible chromosomal combinations in the zygote, disregarding crossing-over, is $(2^n)^2$, and crossing-over greatly increases the number of possible variants. Natural selection has usually favored sexual reproduction because it produces organisms with varied genetic endowments and hence potential for varied adaptation. This variety increases the chances of surviving and reproducing in a changeable environment. Asexual reproduction may be advantageous in a stable environment because it preserves a successful genetic combination.

QUESTIONS

Testing recall

Match each statement below with the stage in mitosis in which it occurs.

a anaphase   d prophase
b interphase   e telophase
c metaphase

1 Aster microtubules appear.
2 Chromosomes migrate to poles.
3 Centromeres uncouple.
4 Centrioles move to poles.
5 Chromosomes reach poles.
6 G_1, S, G_2 stages occur.
7 Nuclear membrane and nucleoli disappear.
8 Chromosomal replication occurs.

Indicate whether each of the following occurs in (a) mitosis, (b) meiosis I, or (c) meiosis II. There may be more than one answer.

9 Double-stranded chromosomes move to poles.
10 Chromosomes shorten and thicken, are double-stranded.
11 Centromeres uncouple.
12 Single-stranded chromosomes move to poles.
13 Nuclear membrane and nucleoli disappear.
14 Haploid cells are produced.
15 The process may produce haploid cells called spores that are specialized for asexual reproduction.
16 Cells genetically identical to the parent cell are produced.
17 Synapsis occurs.

Testing knowledge and understanding

True or false. If false, correct the statement.

1 Much meiosis occurs in the meristematic regions of plants.
2 Centromeres do not uncouple during meiosis I.
3 Duplication of the genetic material (i.e. the synthesis of new genetic material) occurs during interphase.
4 If a given diploid organism has eight chromosomes, four inherited from its mother and four from its father, meiosis will always separate the four maternal chromosomes from the four paternal ones.
5 A cell in prophase I of meiosis has half as many chromosomes as a cell in prophase II.
6 The cells produced by meiosis in plants are not usually specialized as gametes.
7 The vesicles of the cell plate formed during cytokinesis of a plant cell give rise to the middle lamella and the beginnings of the primary cell wall.
8 Mitosis is sometimes complete before cytokinesis begins.

Choose the one best answer.

Questions 9-11 deal with a cell initially containing 24 chromosomes which undergoes one complete meiotic sequence (meiosis I and meiosis II). Use the following answers for these questions; a given answer may be used more than once or not at all.

a 0   d 24
b 6   e 48
c 12

9 How many *chromosomes* are present in each cell in anaphase of meiosis II?
10 How many *homologous pairs of chromosomes* are present in each cell in anaphase of meiosis II?
11 Crossing-over occurs at the stage where the number of *chromatids* in each cell is
12 During the S phase of the cell cycle, the cell

    a undergoes cytokinesis.
    b undergoes meiosis.
    c replicates its DNA.
    d undergoes mitosis.
    e enters interphase.

13 From each primary oocyte that undergoes meiosis, the number of functional egg cells produced is

a one.
b two.
c three.
d four.
e eight.

14 This figure shows a cell whose diploid chromosome number is four. This cell is in

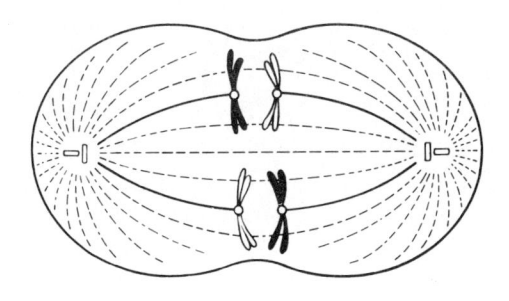

a metaphase.
b anaphase of mitosis.
c anaphase of meiosis I.
d anaphase of meiosis II.
e telophase of mitosis.

15 The above cell is

a plant.
b animal.
c plant or animal.
d bacterial.

16 Suppose the diploid chromosome number of a particular organism is 8. How many different chromosomal combinations could be produced by meiosis in this organism (i.e. how many different kinds of gametes could be formed)? (Exclude complications resulting from crossing-over.)

a 4
b 8
c 16
d 64
e 256

17 Which one of the following is *not* true of meiosis?

a DNA is replicated between each cell division.
b Each chromosome is double during prophase.
c Each chromosome pairs with its homolog during meiosis I.
d Cell division follows chromosome migration.
e Each chromosome may exchange a part of a chromosome with the equivalent part of a homologous chromosome.

For further thought

1 In a single sentence, how would you explain the basic significance of mitosis?

2 Explain the difference between the terms chromosome and chromatid.

3 Which of the phases of mitosis do you think would require the greatest expenditure of energy? Explain your answer.

4 Suggest an hypothesis to explain the adaptive value of having genetic material packaged in chromosomes.

5 In a single sentence, state the fundamental difference in chromosomal behavior between mitosis and meiosis.

6 Without looking at your notes or textbook, take a sheet of paper and make diagrams of mitosis and meiosis in an organism with three pairs of chromosomes. Make a list of the most important differences between mitosis and meiosis.

Chapter 14

PATTERNS OF INHERITANCE

KEY CONCEPTS

1. A gene is a location (locus) on the chromosome that exerts control over one or more characteristics of an organism.

2. In each diploid individual, the genes appear in pairs; when gametes are formed the members of each pair separate and pass into different gametes, so that each gamete has only one of each type of gene.

3. Genes do not alter one another; they remain distinct and segregate unchanged when meiosis occurs.

4. When two or more pairs of genes located on different chromosomes are involved in a cross, the members of one pair are inherited independently of the other.

5. Characteristics are often determined by many genes acting together.

6. The expression of a gene depends both on the other genes present and on the physical environment. All organisms are products of their inheritance *and* their environment.

7. The inheritance patterns for characteristics controlled by genes on the X chromosome are quite different from those whose characteristics are controlled by autosomal genes.

OBJECTIVES

After studying this chapter and reflecting on it, you should be able to

1. Define and use correctly the terms allele, segregation, F_1, F_2.

2. Distinguish between the following pairs of terms: dominant, recessive; homozygous, heterozygous; monohybrid cross, dihybrid cross; phenotype, genotype.

3 Explain how a test cross is performed, and why it is a useful genetic tool.

4 Use a Punnett square or probabilities to do monohybrid and dihybrid crosses.

5 Explain how intermediate inheritance differs from complete dominance.

6 Give the characteristic phenotypic ratios of the F_2 in a dihybrid cross in which the two genes are independent. Explain why complementary genes, epistasis, collaboration, and modifier genes alter this ratio.

7 State Mendel's first and second laws and relate them to the chromosomal theory of inheritance.

8 Distinguish between multiple alleles and multiple gene inheritance.

9 Differentiate between penetrance and expressivity.

10 Explain why statistical analysis is a useful genetic tool.

11 Give the possible genotypes of blood types A, B, AB, and O, and explain the importance of these blood types in giving blood transfusions.

12 Briefly discuss the inheritance of the Rh factor and explain why there is a potential danger to an Rh-positive fetus when the mother is Rh-negative.

13 Explain, using an example, how lethal genes can persist in a population.

14 Give the method of sex determination in *Drosophila* and in human beings.

15 Distinguish between the sex chromosomes and autosomes.

16 Discuss the pattern of inheritance of sex-linked characteristics and explain why recessive sex-linked characteristics are expressed more often in males than in females.

17 Explain what a Barr body is and how it affects gene expression in the female.

18 Distinguish between sex-linked and sex-influenced characteristics.

19 Define linkage, and explain how crossing-over creates genetic variability.

20 Show how crossing-over frequencies are calculated and how they can be used to make chromosomal maps. Compare the map formulated from crossing-over data on *Drosophila* with the map derived from studying the giant chromosomes.

21 Explain how translocation, deletion, duplication, inversion, and nondisjunction alter chromosomes.

22 Do all the problems on pages 621–624 of your textbook and other similar problems posed by your instructor.

SUMMARY

From his breeding experiments on the garden pea the Austrian monk Gregor Mendel concluded that each pea plant possesses two hereditary factors for each character, and that when gametes are formed the two factors segregate into separate gametes. Each new plant thus receives one factor for each character from each parent. The hereditary factors exist as distinct entities within the cell; they do not blend or alter each other, and they segregate unchanged when gametes are formed. This theory is now supported by the occurrence of chromosomal segregation during meiosis.

Mendel was working with two different forms (*alleles*) of the genes for flower color. When both alleles were present, the gene for red flowers was expressed (*dominant*) while the gene for white was masked (*recessive*). The dominant allele is customarily represented by a capital letter, the recessive by a small letter. A diploid cell may be *homozygous* (have two doses of the same allele: C/C, c/c) or *heterozygous* (have one each of two different alleles: C/c). The term *genotype* refers to the possible genetic combinations, *phenotype* to the possible appearances. An easy way to figure out the possible genotypes produced in a cross is to construct a Punnett square.

Extensive investigation has demonstrated that Mendel's results have general validity. Whenever a monohybrid cross is made between two contrasting homozygous individuals, the expected genotype ratio in the second generation of offspring (the F_2) is 1 : 2 : 1. When dominance is involved, the expected phenotypic ratio is 3 : 1.

A test cross is performed to distinguish between homozygous dominant and heterozygous individuals. The unknown is crossed with a homozygous recessive individual. If all the progeny show the dominant phenotype, the unknown genotype is probably homozygous dominant; if any of the progeny show the recessive genotype, the unknown is heterozygous.

One allele is not always completely dominant over the other; in some cases heterozygous individuals show the effects of both elleles and are clearly distinguishable from both homozygous parents. Crosses between heterozygous individuals result in a phenotypic and genotypic ratio of 1 : 2 : 1. This type of inheritance is termed *intermediate inheritance*.

Mendel also made crosses involving two characteristics (a *dihybrid* cross). In his experiments Mendel found that the offspring consistently conformed to a 9 : 3 : 3 : 1 phenotypic ratio. This ratio is characteristic of the F_2 generation of a dihybrid cross (with dominance) in which the genes for the two characters are inherited *independently* of one another.

Some dihybrid crosses involve two independent genes that exert their phenotypic effect on the same character. The ratios from such crosses often differ from the basic ratio. *Complementary genes* are mutually dependent; neither one can exert its phenotypic effect unless the other one does also. *Epistasis* occurs when one gene masks the phenotypic effect of another entirely different gene. Sometimes two different genes interact to produce single-character phenotypes that neither gene alone could produce; this is termed *collaboration*.

Probably no inherited characteristic is controlled by only one gene pair. Even when only one principal gene is involved, other genes act as *modifiers* to influence its expression.

Many characteristics vary in a continuous fashion. This is probably due to *multiple gene inheritance*, in which two or more separate genes affect the same character in an additive fashion.

The action of any gene can be fully understood only in terms of the overall genetic makeup of the individual organism. The expression of a gene depends on the other genes present and on the physical environment. *Penetrance* is the percentage of individuals carrying a gene that actually express it phenotypically. *Expressivity* is the intensity of expression.

In making genetic crosses, ratios are useful in predicting the expected results, but chance alone can cause deviations from the expected. Scientists use statistical analyses to determine whether the deviations they observe in their experimental results are significant. Tests of significance must take into account both the amount of deviation and the sample size, since marked departures from predicted results are not unusual in small samples. Experimental data may be evaluated by a variety of tests, all of which are simply ways to calculate the probability that the deviations are due to chance alone.

Multiple alleles Genes may exist in a number of allelic forms (*multiple alleles*). Each individual has only two alleles for a given trait, but other alleles may be present in the population. An example of multiple alleles in humans is the A-B-O blood series, which involves three alleles: I^A, I^B, and i. Both I^A and I^B are dominant over i, but neither I^A nor I^B is dominant over the other. Individuals with blood type A could be I^A/I^A or I^A/i; type B, I^B/I^B or I^B/i; type AB, I^A/I^B; and type O, ii. Individuals possessing the I^A gene will have antigen A on their red blood cells, those with I^B will have antigen B, those with ii will have no antigen. The two cellular antigens, A and B, react with certain antibodies, anti-A and anti-B, that may be present in the blood. Each antibody will clump red cells containing the corresponding antigen. The presence of these antigens and antibodies has important implications for blood transfusions.

Human Rh blood factors are another example of multiple alleles. The Rh series includes at least nine different antigens. Individuals can be divided into two phenotypic classes, Rh-positive and Rh-negative. Rh-positive individuals (genotype: Rh/Rh or Rh/rh) have the Rh antigen on their red blood cells; Rh-negative individuals (rh/rh) do not. The blood of a normal Rh-negative person will contain no Rh antibody, unless the person is sensitized to the Rh antigen by exposure to it. Once sensitized, the plasma cells will begin synthesizing antibodies.

The Rh factor is important in pregnancies where an Rh-negative mother is carrying an Rh-positive fetus. If any blood seeps between the two circulatory systems the mother may become sensitized and begin producing antibodies. In subsequent pregnancies these may enter the fetal circulation and clump the fetal red cells, which might be fatal to the baby.

Mutations and deleterious genes A variety of influences can cause changes or *mutations* in the chemical structure of genes. Mutations occur constantly; most are deleterious. Natural selection can act against a deleterious gene only if it is expressed phenotypically. Harmful dominant genes can be eliminated rapidly by natural selection, but recessive genes are not expressed in the heterozygous state and thus cannot be easily eliminated.

An allele whose phenotype, when expressed, results in the death of the organism is called a *lethal*. Some alleles that are harmful when homozygous are beneficial when heterozygous. For example, individuals homozygous for the sickle-cell anemia gene suffer from this fatal disease. Heterozygous individuals do suffer from a mild anemia, but they have much higher than normal resistance to malaria. Thus the gene is beneficial in areas where malaria is common. Genes like this, with more than one effect, are said to be *pleiotropic*.

Marriages between closely related individuals are dangerous because they increase the chances of having children who are homozygous for deleterious traits.

Sex and inheritance In most higher organisms where the sexes are separate, the chromosomal endowments of males and females are different. One chromosomal pair, the *sex chromosomes*, differs in size and shape and determines the sex of the individual. All other chromosomes are called *autosomes*. In *Drosophila* and in human beings the females have two large *X chromosomes* whereas the males have an X and a smaller *Y chromosome*. The egg cells produced by meiosis are alike in chromosomal content, but the sperm cells are of two different types, one bearing an X, the other a Y. The sex is determined at the time of fertilization by the type of sperm fertilizing the egg. (The system is reversed in some organisms.)

The genes on the X chromosome are said to be *sex-linked*. Because females have two X chromosomes, they will always have two alleles for any sex-linked characteristic, whereas males will have only one (one on the X, none on the Y). Consequently, recessive sex-linked genes are always expressed phenotypically in the male but may be masked by the dominant allele in the female. In the male, all sex-linked characteristics are inherited from the mother. The Y chromosome apparently has few genes; these genes are said to be *holandric*.

Although females have two X chromosomes in each cell, only one is active; the other coils up into a tiny dark *Barr body* whose genes are inactive.

Many genes that control sexual characteristics are located on the autosomes of both sexes; these should not be confused with sex-linked characteristics.

Linkage Genes that are located on different chromosomes can segregate independently of one another during meiosis. Genes located on the same chromosome are *linked* together; they ordinarily remain together during meiosis. However, crossing-over can occur between homologous chromosomes during synapsis in meiosis I; this results in new linkages.

The frequency of crossing-over between any two linked genes will be proportional to the distance between them. The percentage of crossing-over can be used to map gene locations. By convention, one unit of map distance is the distance within which crossing-over occurs one percent of the time. Crossing-over is the classical test of whether two characters are controlled by one gene or by two separate genes.

The giant chromosomes that occur in the salivary glands of many flies have been useful in determining visually the location of individual genes on the chromosomes. Such maps show that sequences of the genes determined by crossing-over frequencies are correct but the distances between the genes are not.

Chromosomal alterations Crossing-over is one kind of chromosomal rearrangement; other alterations occur as well. In *translocation*, portions of two nonhomologous chromosomes are exchanged. Sometimes a portion of a chromosome breaks off and is lost, resulting in a *deletion*. Sometimes a piece breaks off one chromosome and fuses onto the end of its homologous chromosome, forming a *duplication*. Occasionally a portion of a chromosome breaks out, turns around, and fuses in its original position in reversed order, forming an *inversion*. This may alter gene expression because of the *position effect*.

Separation of the chromosomes during meiosis does not always occur normally; sometimes both members of one homologous pair move to the same pole. The result of this *nondisjunction* may be the production of a cell with an extra chromosome (*trisomy*). Occasionally all the chromosomes move to the same pole. If this cell then unites with another during fertilization, the resulting zygote has more than two sets of chromosomes and is said to be *polyploid*.

Study guide for genetics problems

The best way to gain an understanding of genetics is to work with it. The problems at the end of Chapter 14 in your text illustrate the various patterns of inheritance the chapter discusses. The following information is intended to get you started on the problems. Answers are provided in this section, but it is strongly advised that you work the problems *first* and check your answers later. Genetics problems are considerably easier when you start with the answer and work backwards, but you will not learn how to do problems that way.

General rules

1 Know your terms: dominant (capital letter), recessive (small letter); P, F_1, F_2; homozygous, heterozygous; genotype, phenotype.

2 Read the problem and assign appropriate symbols to the characteristics.

3 Set up a Punnett square correctly: female gametes go along the left side, male gametes across the top. Be sure to use all possible gametes.

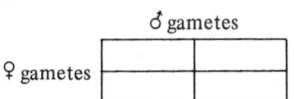

	♂ gametes	
♀ gametes		

4 Instead of a Punnett square, use probability to solve problems. The basic principle is that the chance that a number of independent events will occur together equals the chance of the first event times the chance of each following event.

Example: The probability of getting heads when flipping a coin is ½ (or 0.5). If two pennies are flipped at the same time, the chance of getting two heads is the chance of the first being heads (½) times the chance of the second (½) or (½) X (½) = ¼. The chance of getting five heads in a row is (½) X (½) X (½) X (½) X (½) = $1/32$. (See question 19.)

Types of problems

1 Monohybrid cross: cross involving one character

Example: If pole bean plants (*P*) are dominant to bush beans (*p*), what are the expected genotypes and phenotypes when two heterozygous individuals are mated?

P   female X male
     P/p   P/p

gametes: P and p P and p

Punnett square:

	♂ P	p
♀ P	P/P	P/p
p	P/p	p/p

genotype ratio: 1 P/P : 2 P/p : 1 p/p

phenotype ratio: 3 pole : 1 bush

(Now try questions 1–3 in your text.)

2 Test cross: unknown organism is crossed with a homozygous recessive.

Example: If the genotype of a pole bean plant is in doubt (*P/–*), it is crossed with a bush bean plant.

    P P/– X p/p

a If – is dominant (*P*) the cross is P/P X p/p and all offspring are pole.

b If – is recessive (*p*) the cross is P/p X p/p and half the offspring are pole and half bush.

      P/p X p/p

gametes: P, p   p

Punnett square:

	p
P	P/p
p	p/p

genotype ratio: 1 P/p : 1 p/p

phenotype ratio: 1 pole : 1 bush

(See question 2.)

3 Dihybrid cross: cross involving two characters

Example: Plants heterozygous for red pole beans were crossed with each other. The genes for color (R = red, r = white) and shape (P = pole, p = bush) are on different chromosomes.

a Solving the problem using a Punnett square.

	female		male
parents:	R/r P/p	×	R/r P/p
gametes:	RP, Rp, rP, rp	×	RP, Rp, rP, rp

Punnett square:

	RP	Rp	rP	rp
RP	R/R P/P	R/R P/p		
Rp				
rP				
rp				

Complete the Punnett square and determine the genotype and phenotype ratios. (Now try question 8.)

b Solving the problem using probabilities.

It is very laborious to make Punnett squares for two or more pairs of genes, so probability is often used instead. A convenient and rapid way to do problems where you are asked to calculate all the genotype and phenotype ratios is to use probabilities and the fork-line method to do crosses.

According to Mendel's second law, the members of one pair of alleles are inherited independently of the members of another. Consequently, the probability of each of the various genotypic combinations in the offspring is equal to the product of the genotypic probabilities of each pair of alleles.

So, using the fork-line method, one calculates the genotypic probabilities of each pair of alleles separately, and then finds all possible genotypic combinations and multiplies their separate probabilities.

Example: R/r P/p × R/r P/p

1 Consider each pair of alleles separately.

Rr × Rr gives you 1/4 R/R, 2/4 R/r, 1/4 r/r (3 possible genotypes)

P/p × P/p gives you 1/4 P/P, 2/4 P/p, 1/4 p/p (3 possible genotypes)

(3 × 3 = 9 total possible genotypic combinations)

2 Now find all possible combinations.

Genotypes of offspring

1/4 R/R
— 1/4 P/P = 1/16 R/R P/P
— 2/4 P/p = 2/16 R/R P/p
— 1/4 p/p = 1/16 R/R p/p

2/4 R/r
— 1/4 P/P = 2/16 R/r P/P
— 2/4 P/p = 4/16 R/r P/p
— 1/4 p/p = 2/16 R/r p/p

1/4 r/r
— 1/4 P/P = 1/16 r/r P/P
— 2/4 P/p = 2/16 r/r P/p
— 1/4 p/p = 1/16 r/r p/p

If you make a cross involving three pairs of genes, you can simply add on another column of "forks." (Start with a big piece of paper!)

Probabilities can be used to solve many genetics problems that would be tedious to solve with a Punnett square.

Example: A heterozygous red pole smooth bean plant (R/r P/p S/s) is crossed with a red pole bean plant with wrinkled seeds (R/r P/P s/s). What fraction of the offspring will have the genotype R/r P/P s/s?

the chance of R/r from the cross R/r × R/r is ½

the chance of P/P from the cross P/p × P/P is ½

the chance of s/s from the cross S/s × s/s is ½

Therefore the chance of an R/r P/P s/s is equal to the product of their separate probabilities or (½) × (½) × (½) = ⅛. (Now try question 19.)

4 Sex linkage: Genes for sex-linked characteristics are carried on the X chromosome. Because females have two X chromosomes they will always have two alleles for a particular characteristic whereas males will have only one. Consequently, recessive sex-linked traits will show up more often in males than in females.

Sometimes students find it helpful to use different symbols for sex-linked problems; X^C can represent a dominant sex-linked gene and X^c the recessive. The Y chromosome is designated Y.

Example: Hemophilia is sex-linked and recessive. A normal man marries a woman who is a carrier (heterozygous) for this trait. What phenotypes will the children have?

Let X^H = allele for normal blood
X^h = allele for hemophilia

The cross is ♀ X^H/X^h × ♂ X^H/Y

Results: ½ daughters will be normal
½ will be carriers
½ sons will have hemophilia

(Now try problem 22.)

5 Linkage: Genes are linked when they are on the same chromosome. In making crosses involving linked genes, different symbols are used. If the genes A and B are linked they are represented as follows:

$\dfrac{AB}{ab}$ short for $\dfrac{AB}{ab}$ or

Linked genes are sometimes separated when homologous chromosomes change parts during synapsis in meiosis:

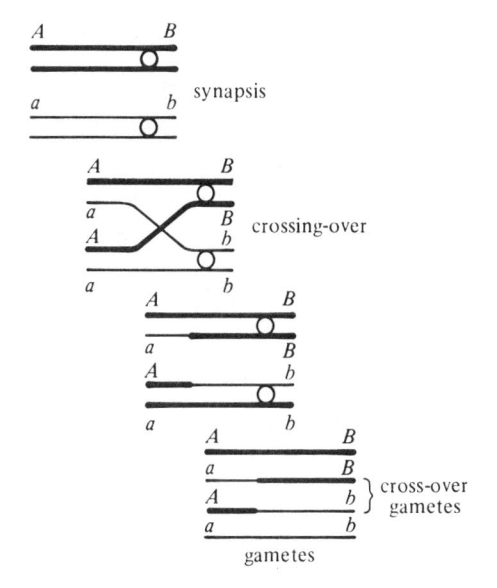

The normal and most common gametes are the noncrossovers, AB and ab. The rare new gametes formed as a result of crossing-over are Ab and aB. Linkage and crossing-over can be detected in a test cross:

$$\dfrac{AB}{ab} \times \dfrac{ab}{ab}$$

	ab	# of offspring
Normal gamete AB	$\dfrac{AB}{ab}$	45
Normal gamete ab	$\dfrac{ab}{ab}$	44
Crossing-over gamete aB	$\dfrac{aB}{ab}$	5
Crossing-over gamete Ab	$\dfrac{Ab}{ab}$	5

There are four phenotypes but they are not in equal proportions. (If A and B were not linked and were on different chromosomes the phenotypes would be 1 : 1 : 1 : 1.)

6 Chromosomal mapping: The distance between two genes can be calculated using the following procedure:

$$\dfrac{\text{number of crossing-over progeny}}{\text{total number of progeny}} \times 100 = \% \text{ crossing-over}$$

1% crossing-over = 1 map unit

From our example above,

$$\dfrac{5+5}{100} \times 100 = 10\% = 10 \text{ map units}$$

(Now try question 29.)

Now do all the rest of the problems starting on page 621 of your textbook, and check your answers. Additional problems are given in the Questions section.

Answers to genetics problems, in text (pp. 621–624)

1 a All heterozygous, white

	w	w
W	W/w	W/w
W	W/w	W/w

b Genotype ratio: W/w : w/w is 1 : 1
Phenotype ratio: 1 white : 1 yellow

	w	w
W	W/w	W/w
w	w/w	w/w

c Genotype ratio: 1 W/W : 2 W/w : 1 w/w
Phenotype ratio: 3 white : 1 yellow

	W	w
W	W/W	W/w
w	W/w	w/w

2 Let Y be the allele for white color and y be the allele for yellow color. Thus, the genotype of the pollen, the male gamete, is Y/y and the female gamete is y/y. The cross is $Y/y \times y/y$, and the offspring, determined by the Punnett square, are as follows: the genotype ratio, $Y/y : y/y$, is 1 : 1; the phenotype ratio, white : yellow, is 1 : 1.

	♂ Y	y
♀ y	Y/y white	y/y yellow
y	Y/y white	y/y yellow

3 Let B be for brown eyes and b be for blue eyes. The man must be b/b, but the woman is $B/-$. However, since her father is b/b, she must have a b allele. Thus, she is B/b. The cross is $b/b \times B/b$, and the Punnett square shows that 50 percent blue eyes are predicted.

	♂ b	b
♀ B	B/b brown	B/b brown
b	b/b blue	b/b blue

4 This time the man is $B/-$, and the woman is b/b. The cross is either $B/b \times b/b$, or $B/B \times b/b$. The first cross would yield 50 percent blue-eyed children as in question 2, and the second cross would yield all brown-eyed children as shown in the figure. Note that every child must receive a dominant B allele from his father. In this case the father is probably homozygous, but one cannot be certain that this is true. Furthermore, an eleventh brown-eyed child makes no difference. It cannot be proved beyond a doubt that a character is homozygous dominant. In this case there is a one in 2^{11} chance that the father is heterozygous, or some other factor may be prohibiting his b allele from being expressed.

	♂ B	B
♀ b	B/b brown	B/b brown
b	B/b brown	B/b brown

5 In this problem the brown-eyed man is B/b because his mother was b/b. His father is either B/B or B/b. His wife is b/b, so her brown-eyed parents must both be heterozygous, B/b. The blue-eyed son is, of course, b/b.

6 There are probably a number of explanations, but it is best to assume only one genetic locus at first. This is probably true since only tail length is involved. The ratio is 3 : 6 : 2, which by inspection is similar to 1 : 2 : 1. This ratio appears as the expected genotype ratio of a monohybrid cross. This would mean that the parental, short-tailed cats are heterozygous, T^1/T^2. Assuming then that T^1/T^1 yields a long tail and T^2/T^2 yields no tail, the Punnett square confirms this possible explanation. This assumes no dominance, with short tails a hybrid between none and long alleles.

	♂ T^1	T^2
♀ T^1	T^1/T^1 long	T^1/T^2 short
T^2	T^1/T^2 short	T^2/T^2 none

7 Again, assume that only a single locus is involved. The ratio of pups (haired : hairless : deformed) is 1 : 2 : ? . This is most like a monohybrid cross ratio, 1 : 2 : 1. This implies that the hairless dogs are monohybrids, H^1/H^2, which is substantiated by the number of hairless pups relative to haired pups, who would be H^1/H^1. The H^2 allele is then assumed to be lethal in the homozygous condition. A normal hairless cross, $H^1/H^1 \times H^1/H^2$, would yield $H^1/H^1 : H^1/H^2$ in the ratio 1 : 1, in agreement with the first statement.

8 *a* This is a dihybrid cross, as in questions 6 and 9. Therefore the genotype ratio is $T/T\ S/S : T/T\ S/s : T/T\ s/s : T/t\ S/S : T/t\ S/s : T/t\ s/s : t/t\ S/S : t/t\ S/s : t/t\ s/s$– in the ratio 1 : 2 : 1 : 2 : 4 : 2 : 1 : 2 : 1. The phenotypes are tall, smooth : tall, wrinkled : short, smooth : short, wrinkled in the ratio 9 : 3 : 3 : 1.

b This is a test cross and the phenotypes will reflect the alleles present in the tall, wrinkled parent in equal numbers; tall, wrinkled : short, wrinkled–1 : 1. The genotypes are $T/t\ s/s : t/t\ s/s$–1 : 1. Note that there can be no homozygous dominant in a test cross.

c In this case each parent has only two possible gametes, *tS* and *ts*, and *Ts* and *ts*, respectively. The genotypes are therefore *T/t S/s : T/t s/s : t/t S/s : t/t s/s*–1 : 1 : 1 : 1. The phenotypes are tall, smooth : tall, wrinkled : short, smooth : short, wrinkled–1 : 1 : 1 : 1.

d Here each parent has only one possible gamete; thus there is only one possible offspring, *T/t S/s*, tall and smooth.

9 Let *B* be belted, *b* be no belt, *F* be fused, and *f* be normal. The cross is *b/b F/F* × *B/B f/f*, yielding all double heterozygotes, *B/b F/f*, in the F_1 generation. The Punnett square shows the results of freely interbreeding these individuals, and the genotype and phenotype ratios are again 1 : 2 : 1 : 2 : 4 : 2 : 1 : 2 : 1 and 9 : 3 : 3 : 1.

	BF	Bf	bF	bf
BF	B/B F/F belted fused	B/B F/f belted fused	B/b F/F belted fused	B/b F/f belted fused
Bf	B/B F/f belted fused	B/B f/f belted norm	B/b F/f belted fused	B/b f/f belted norm
bF	B/b F/F belted fused	B/b F/f belted fused	b/b F/F even fused	b/b F/f even fused
bf	B/b F/f belted fused	B/b f/f belted norm	b/b F/f even fused	b/b f/f even norm

10 Let *S* be green, *s* be striped, *L* be short, and *l* be long. The cross is *s/s l/l* × *S/s L/l*. Since this is a test cross the phenotypes will reflect the alleles present in the green, short parent plant in equal numbers. (Assume no linkage.) Thus green, short : green, long : striped, short : striped, long–1 : 1 : 1 : 1.

11 Let *S* be long-winged, *s* be short-winged, *H* be hairless, and *h* be hairy. The parental cross is *s/s h/h* × *S/S H/H*. Since the only gametes are *s/h* and *S/H*, the only F_1 offspring are *S/s H/h*, double heterozygotes. The Punnett square shows the results of the F_2 generation. The genotypic ratio will be *S/S H/H : S/S H/h : S/S h/h : S/H H/H : S/s H/h : S/s h/h : s/s H/H : s/s H/h : s/s h/h*–1 : 2 : 1 : 2 : 4 : 2 : 1 : 2 : 1, and the phenotypic ratio will be long-winged, hairless : long-winged, hairy : short-winged, hairless : short-winged, hairy–9 : 3 : 3 : 1. The F_1 phenotypes are long-winged, hairless.

	SH ♂	Sh	sH	sh
SH	S/S H/H long, less	S/S H/h long, less	S/s H/H long, less	S/s H/s long, less
Sh ♀	S/S H/h long, less	S/S h/h long, hairy	S/s H/h long, less	S/s h/h long, hairy
sH	S/s H/H long, less	S/s H/h long, less	s/s H/H short, less	s/s H/h short, less
sh	S/s H/h long, less	S/s h/h long, hairy	s/s H/h short, less	s/s h/h short, hairy

12 The cross in this case is *s/s H/h* × *S/s h/h*. The male can only contribute gametes with *s/H* and *s/h* while the female can contribute *S/h* and *s/h*. The Punnett square can be set up accordingly, indicating the following phenotypes: long-winged, hairless : short-winged, hairless : long-winged, hairy : short-winged, hairy–1 : 1 : 1 : 1.

	sH ♂	sh
Sh ♀	S/s H/h long, less	S/s h/h long, hairy
sh	s/s H/h short, less	s/s h/h short, hairy

13 Let *S* be barking, *s* be silent, *D* be erect, and *d* be drooping. The breeder wants *SSdd*. Any droop-eared dog must be pure and thus presents no problem. However, the breeder must use a test cross on his barkers since they may be heterozygous: *S/- d/d* × *s/s -/-*. (The ears of the test dog are irrelevant.) Referring to problem 4, if the dog is homozygous, as the breeder wishes, all pups will be *S/s*, barkers, but if it is heterozygous, about 50 percent should be silent. However, as stated in problem 4, he can never be sure of having a homozygous dominant trait.

14 The cross is *A/- R/-* × *A/- r/r*. If the male's second black allele were *R*, then there could be no yellow or cream offspring. Thus the allele is *r*. Similarly, if either second yellow allele were *A*, there could be no black or cream offspring. Thus they are both *a*. The cross was *A/a R/r* × *A/a r/r*.

126 • CHAPTER 14

15 The cross is $A/\text{-} \; r/r \times a/a \; R/\text{-}$, and there are no black or cream offspring. Thus the second yellow allele of the male is not likely to be an a, which would make 50 percent black expected. The second black allele of the female must be recessive, r, to produce yellow offspring. The cross was $A/A \; r/r \times a/a \; R/r$.

16 There are two possible crosses, $C/\text{-} \; i/i \times C/\text{-} \; I/\text{-}$ and $C/\text{-} \; i/i \times c/c \; \text{-}/\text{-}$. At least half the offspring of the first cross would receive an I allele and be colorless, but all the offspring are colored. Using the same argument in the second cross, neither inhibitory allele of the hen can be I. Finally, if the cock's second color allele were c, one would expect 50 percent white. Therefore, $C/C \; i/i \times c/c \; i/i$ is the most likely cross. The offspring are all $C/c \; i/i$.

17 The Punnett square shows the two gametes contributed by the parents. It shows that 75 percent of their offspring will be deaf. Without using a Punnett square one could note that 50 percent of the offspring would have the necessary K allele for hearing, but only 50 percent of these, and thus 25 percent of the total, would have it without the nasty M allele.

	♂ kM	km
♀ Km	$K/k \; M/m$ deaf	$K/k \; m/m$ normal
km	$k/k \; M/m$ deaf	$k/k \; m/m$ deaf

18 Assuming no linkage, each character, Ks, Ls, and Ms, assorts independently. The chances of any specific homozygous recessive appearing is 1/4. The chance of all three appearing is therefore (1/4)(1/4)(1/4) or 1/64.

19 The logic is the same here, where the parents are heterozygotes in five traits. In addition, the probability of any one heterozygous combination appearing is 1/2. Thus, the chance of this particular offspring is (1/4)(1/4)(1/2)(1/4)(1/2) or 1/256.

20 The man is I^B/i since one of his parents is ii, which is actually irrelevant. The cross is $I^B/i \times I^A/I^B$, so that half of the offspring will have an I^A allele from the mother. They cannot be pure B type. The other half will get the I^B and will be B.

21 Every parent has one unknown allele and thus Shirley, who is i/i, may belong to either family. Jane, who is I^B/i, cannot be the daughter of the Joneses, where no I^B allele is present. Thus, a mixup did occur in her case, and she could belong to the Smiths. However, since Shirley may also be their daughter, a third family may be involved.

22 The cross is $X^B/Y \times X^B/X^b$. As a result ½ of the male offspring will receive the lethal allele and die. All the females will receive the X^B from their father and survive. Therefore, there will be twice as many female as male children.

23 Let C be normal and c be color-blind. The cross is $X^C/Y \times X^c/X^c$. Thus all males receive an X^c from their mother and will be color-blind. All the females will receive the dominant X^C from their father and be normal.

24 Yes, genetics can, and the man does have grounds. Since he is normal, X^C/Y, he will give all his daughters the dominant allele for the normal condition. (Perhaps the girl was switched in the hospital.)

25 As in question 7, a lethal gene is probably involved. Since a disproportionate number of males survived, the allele is no doubt sex-linked. In fact, in accordance with the data given, ½ of the female Z/W offspring died. If the lethal allele were on the W chromosome, all the female offspring would die. The only other possibility is that it is heterozygous in the male, Z^L/Z. The cross would be $Z^L/Z \times Z/W$, which would yield all viable males and 50 percent viable females (1/3 : 2/3).

26 Let B be for barred and b be for nonbarred. The male cannot be barred. If he is homozygous, all offspring will be barred, and if he is heterozygous, only ½ of the females will be barred in the F_1. If the hen is barred, Z^B/W, and the cock is not, Z^b/Z^b, then all offspring cocks will be Z^B/Z^b, barred, and the hens will all be Z^b/W, nonbarred. This, of course, only works for the F_1 generation, since the cocks are now barred.

27 Let L be short hair and l be long hair and X^{B1} be yellow and X^{B2} black. The cross is $l/l\ X^{B2}/Y \times L/L\ X^{B1}/X^{B2}$. All the kittens will be L/l, short-haired. The color ratios are $X^{B1}/Y : X^{B2}/Y : X^{B1}/X^{B2} : X^{B2}/X^{B2}$—yellow male : black male : tortoise female : black female–1 : 1 : 1 : 1. If these freely interbreed, ¼ will be long-haired by the monohybrid ratio. Half of these will of course be males, and only ¼ of the males will be yellow since one out of the four female alleles is X^{B1}. Thus the chance of a long-haired, yellow male is (1/4)(1/2)(1/4) or 1/32. The Punnett square illustrates this. Two out of 64 possibilities are such a cat. Each 4 × 4 block sector represents one of the four possible crosses. The two long-haired yellow males have thicker lines around them.

♀ \ ♂	LX^1	LY	lX^1	lY	LX^2	LY	lX^2	lY
LX^1	$L/L\ X^1/X^1$ s, y, f	$L/L\ X^1/Y$ s, y, m	$L/l\ X^1/X^1$ s, y, f	$L/l\ X^1/Y$ s, y, m	$L/L\ X^2/X^1$ s, t, f	$L/L\ X^1/Y$ s, y, m	$L/l\ X^2/X^1$ s, t, f	$L/l\ X^1/Y$ s, y, m
LX^2	$L/L\ X^1/X^2$ s, t, f	$L/L\ X^2/Y$ s, b, m	$L/l\ X^1/X^2$ s, t, f	$L/l\ X^2/Y$ s, b, m	$L/L\ X^2/X^2$ s, b, f	$L/L\ X^2/Y$ s, b, m	$L/l\ X^2/X^2$ s, b, f	$L/l\ X^2/Y$ s, b, m
lX^1	$L/l\ X^1/X^1$ s, y, f	$L/l\ X^1/Y$ s, y, m	$l/l\ X^1/X^1$ l, y, f	$l/l\ X^1/Y$ l, y, m	$L/l\ X^2/X^1$ s, t, f	$L/l\ X^1/Y$ s, y, m	$l/l\ X^2/X^1$ l, t, f	$l/l\ X^1/Y$ l, y, m
lX^2	$L/l\ X^1/X^2$ s, t, f	$L/l\ X^2/Y$ s, b, f	$l/l\ X^1/X^2$ l, t, f	$l/l\ X^2/Y$ l, b, m	$L/l\ X^2/X^2$ s, b, f	$L/l\ X^2/Y$ s, b, m	$l/l\ X^2/X^2$ l, b, f	$l/l\ X^2/Y$ l, b, m
LX^2	$L/L\ X^1/X^2$ s, t, f	$L/L\ X^2/Y$ s, b, m	$L/l\ X^1/X^2$ s, t, f	$L/l\ X^2/Y$ s, b, m	$L/L\ X^2/X^2$ s, b, f	$L/L\ X^2/Y$ s, b, m	$L/l\ X^2/X^2$ s, b, f	$L/l\ X^2/Y$ s, b, m
LX^2	$L/L\ X^1/X^2$ s, t, f	$L/L\ X^2/Y$ s, b, m	$L/l\ X^1/X^2$ s, t, f	$L/l\ X^2/Y$ s, b, m	$L/L\ X^2/X^2$ s, b, f	$L/L\ X^2/Y$ s, b, m	$L/l\ X^2/X^2$ s, b, f	$L/l\ X^2/Y$ s, b, m
lX^2	$L/l\ X^1/X^2$ s, t, f	$L/l\ X^2/Y$ s, b, m	$l/l\ X^1/X^2$ l, t, f	$l/l\ X^2/Y$ l, b, m	$L/l\ X^2/X^2$ s, b, f	$L/l\ X^2/Y$ s, b, m	$l/l\ X^2/X^2$ l, b, f	$l/l\ X^2/Y$ l, b, m
lX^2	$L/l\ X^1/X^2$ s, t, f	$L/l\ X^2/Y$ s, b, m	$l/l\ X^1/X^2$ l, t, f	$l/l\ X^2/Y$ l, b, m	$L/l\ X^2/X^2$ s, b, f	$l/l\ X^2/Y$ s, b, m	$l/l\ X^2/X^2$ l, b, f	$l/l\ X^2/Y$ l, b, m

28 Number the individuals in the diagram 1 through 15 from left to right and top to bottom.

a The trait is not dominant autosomal because female 9 is deaf while neither parent is.
b The trait could be recessive autosomal.
c It is not sex-linked dominant for the same reason as a.
d It is not sex-linked recessive, for all the sons of female 2 would have to be deaf and male 8 is not. Female 9 could not be deaf since her father, male 3, would have to be deaf for her to be homozygous recessive.
e The trait is not holandric, for no female could have the trait. It would also have to appear in males 1 and 8.

29 Let B be gray, b be black, V be normal, and v be vestigial. The cross is

$$\frac{BV}{BV} \times \frac{bv}{bv}$$

yielding all

$$\frac{BV}{bv}$$

The next cross

$$\frac{BV}{bv} \times \frac{bv}{bv}$$

a test cross, should segregate the characters into four equally occurring phenotypes if they are not linked. Since the phenotypes do not occur equally, the genes are linked and crossing-over occurred 111/600 or 18.5 percent of the time. Therefore, the characters are 18.5 map units apart.

30 The genes are linked since a ratio of 1 : 1 : 1 : 1 would be expected if the genes were inherited independently.

% crossover = $\frac{14 + 10}{200}$ = 12% = 12 map units apart

31 If the genes are 20 units apart, crossing-over would occur 20 percent of the time. Thus 20 percent of the 1,000 offspring would be expected to show the crossover phenotypes. (1000 × .20 = 200)

32 If A and B are 40 map units apart, and both A and B are 20 units from C, then C must lie halfway between them. Similarly, since both C and B are 10 units from D, D must lie halfway between them. The diagram below illustrates the conclusion through the elimination of the various possible alternatives in the order given in the text. The individual steps are lettered.

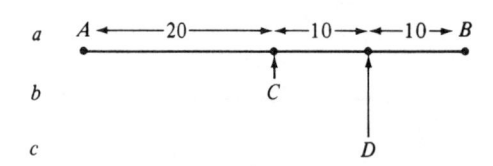

a A ←——20——→ ←—10—→ ←—10→ B
b   C
c   D

33

Genotype	Drosophila sex	Human sex	Number of Barr bodies
XO	male	female	none
XXX	female	female	two
XYY	male	male	none
XXXX	female	female	three
XXXY	female	male	two
XXXXY	female	male	three

QUESTIONS

Choose the one best answer.

1 According to Mendel's *second* law, the Law of Independent Assortment,
   a the two "factors" or genes influencing a certain trait separate during gamete formation.
   b maternal and paternal traits are blended in the offspring.
   c each gene is inherited separately from other genes.
   d one gene is dominant to another.

2 Pleiotropism describes
   a a single gene having multiple effects.
   b gene interaction of multiple alleles.
   c a single trait being influenced by several genes.
   d a trait that is not expressed for several generations.
   e polygenic inheritance.

3 In pigeons, the grizzle color pattern depends on a dominant autosomal gene G. A mating of two grizzle birds produced one nongrizzle youngster this year. If this pair of pigeons produces more youngsters next year, what percentage would be expected to be grizzles?

   a 100 percent   d 25 percent
   b 75 percent   e 0 percent
   c 50 percent

4 Singer Arlo Guthrie has a 50 percent chance of dying prematurely from the same genetic disease that killed his father, Woody Guthrie. Neither Woody Guthrie's mother nor Arlo Guthrie's mother carries any allele for this disease (Huntington's Chorea). What type of inheritance pattern does this disease have? (It can be figured out from the information given.)

   a autosomal dominant
   b sex-linked dominant
   c autosomal recessive
   d sex-linked recessive

5 In cocker spaniels, black color is due to a dominant gene B, and red color to its recessive allele b. Solid color is dependent on a dominant gene S, and white spotting on its recessive allele s. A solid red male was mated to a black-and-white female. They had five puppies: one black, one red, one black-and-white, and two red-and-white. What were the genotypes of the parents?

   a male *bbss* and female *BBss*
   b male *bbSs* and female *Bbss*
   c male *bbSs* and female *BbSs*
   d male *BbSs* and female *Bbss*
   e male *BbSS* and female *Bbss*

6 Ignoring modifier genes, we can think of brown eyes in human beings as determined by a dominant autosomal allele B and blue eyes by a recessive allele b; free earlobes are determined by a dominant autosomal allele F and attached earlobes by a recessive allele f. A brown-eyed man with attached earlobes (whose mother was blue-eyed) marries a blue-eyed woman with free earlobes (whose father had attached earlobes). What phenotypes may be expected among their children?

   a Both blue- and brown-eyed children may be expected, but all will have free earlobes.
   b All four possible combinations of eye color and earlobe condition may be expected, in roughly equal frequencies.
   c All brown-eyed children will have attached earlobes, and all blue-eyed children will have free earlobes.
   d Both blue- and brown-eyed children may be expected, but all will have attached earlobes.
   e All brown-eyed children will have free earlobes, and all blue-eyed children will have attached earlobes.

7 In a diploid sexually reproducing organism with five pairs of chromosomes (I, II, III, IV, and V), what are the chances that one of its I, one of its III, and one of its V chromosomes were inherited from its paternal grandfather? (Ignore complications resulting from crossing-over.)

   a 1/4
   b 1/2
   c 1/8
   d 1/16
   e 1/64

8 Knowledge of the blood-type genotypes of a certain couple leads us to say that if they were to have many children, the ratios of the children's blood types would be expected to approximate ½ type A and ½ type B. It follows that the blood types of the couple are

   a A and B
   b AB and AB
   c AB and B
   d AB and A
   e AB and O

9 An Rh^+ man whose mother was Rh^- marries an Rh^- woman whose father was Rh^+. Assuming the woman has never had a blood transfusion, what are the chances this couple will have an Rh^+ first child who suffers from erythroblastosis fetalis?

   a 0 percent
   b 25 percent
   c 50 percent
   d 75 percent
   e 100 percent

10 A particular sex-linked recessive disease of human beings is usually fatal. Suppose that, by chance, a boy with the disease lives past puberty and marries a woman heterozygous for the trait. If they have a daughter, what is the probability that she will have the disease?

   a 0 percent
   b 25 percent
   c 50 percent
   d 75 percent
   e 100 percent

11 The rare trait ocular albinism (almost complete absence of eye pigment) is inherited as a sex-linked recessive. A man with ocular albinism marries a woman who neither has this condition nor is a carrier. Which one of the following is the best prediction concerning their offspring?

   a All their sons will have ocular albinism, and all their daughters will be carriers.
   b All their children of both sexes will have ocular albinism.
   c About 50 percent of their sons will have ocular albinism, and all their daughters will be carriers.
   d About 50 percent of their daughters will have ocular albinism, but all their sons will have normal eyes.
   e None of their children will have ocular albinism, but all their daughters will be carriers.

12 In a P cross, an $AABBCC$ individual is paired with an $aabbcc$ individual. Assuming no linkage, what will be the expected frequency of $AAbbCc$ individuals in the F_2 generation?

   a 16/64
   b 8/64
   c 4/64
   d 2/64
   e 1/64

In answering questions 13-15, refer to the following pedigree for one type of deafness in human beings. Squares symbolize males, circles females; filled symbols designate deaf individuals, open symbols individuals with normal hearing.

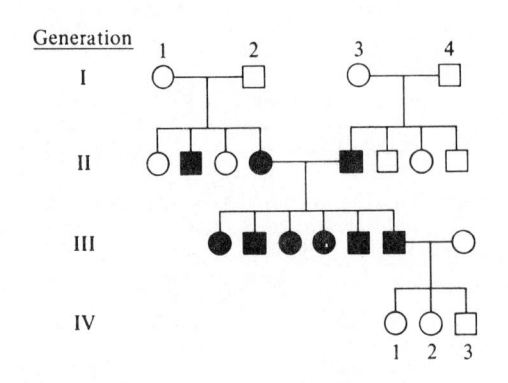

13 This type of deafness is probably inherited as

   a an autosomal dominant.
   b an autosomal recessive.
   c a sex-linked dominant.
   d a sex-linked recessive.
   e a holandric.

14 The genotype of individual 4 in generation I is probably

   a DD
   b Dd
   c dd
   d $X^D Y$
   e $X^d Y$

15 The genotype of individual 3 in generation IV is probably

   a DD
   b Dd
   c dd
   d $X^D Y$
   e $X^d Y$

16 A certain man has the genotype AaBb; genes A and B are on one chromosome, and a and b are on the homologous chromosome. Suppose crossing-over occurs during a meiotic division in this man's testis. How many genetically different (with regard to the two genes discussed here) types of sperm cells will result?

   a one
   b two
   c four
   d eight
   e sixteen

17 The crossing-over frequency between genes A and B is 35 percent; between B and C, 10 percent; between C and D, 15 percent; between C and A, 25 percent; between D and B, 25 percent. The sequence of the genes on the chromosome is

   a ACDB.
   b ACBD.
   c ABDC.
   d ABCD.
   e ADCB.

18 Consider two linked autosomal genes. The dominant allele C of the first gene causes cataracts of the eye, whereas its recessive allele c produces normal eyes. The dominant allele of the second gene P causes polydactyly (presence of an extra finger on each hand), whereas its recessive allele p produces normal hands. A man with cataracts and normal hands marries a woman with polydactyly and normal eyes. Their son has both cataracts and polydactyly. The son marries a woman with neither trait. Assuming no crossing-over, what is the probability that their first child will have both cataracts and polydactyly?

   a 0 percent
   b 25 percent
   c 50 percent
   d 75 percent
   e 100 percent

19 Gene A and Gene B are known to be 10 units apart on the same chromosome. Individuals homozygous dominant for these genes were mated with the homozygous recessives. The offspring were then test-crossed. If there were 1000 offspring from the test cross, how many of the offspring would you predict would show the crossover phenotypes?

   a 10
   b 50
   c 100
   d 250
   e 500

20 The exchange of parts between nonhomologous chromosomes is called

   a inversion.
   b translocation.
   c transduction.
   d transformation.
   e duplication.

21 A person with XYY syndrome will have how many Barr bodies in his nuclei?

   a 0
   b 1
   c 2
   d 3
   e either 1 or 2

Chapter 15

THE NATURE OF THE GENE AND ITS ACTION

KEY CONCEPTS

1. DNA is the genetic material that makes up the genes; it can replicate itself and controls the functioning of all cells.

2. Genes encode the information that directs the synthesis of the enzymes that control the cell's chemical reactions; these reactions determine the phenotypic characteristics.

3. The sequence of the bases in the DNA indicates the sequence in which amino acids must be linked in protein synthesis.

4. The genetic code is a triplet code; a group of three nucleotides specifies one amino acid. The code is degenerate, nonoverlapping, and universal.

5. Some cytoplasmic structures have their own DNA and are capable of self-replication.

6. There is no sharp distinction between bacterial and viral genetics; a piece of DNA may function as a bacterial gene at one time and as a viral gene a little later.

OBJECTIVES

After studying this chapter and reflecting on it, you should be able to

1. Cite two lines of experimental evidence to support the conclusion that DNA is the genetic material.

2. Name the three components of nucleotides and the four nitrogenous bases found in DNA; indicate which are pyrimidines and which are purines.

3. Describe the Watson-Crick model and the evidence on which it was based. Show how this model accounts for precise replication of the genetic material.

4. Give the sequence of a new chain of DNA that is replicated from a strand of DNA with the following sequence of nitrogenous bases: adenine, guanine, cytosine, adenine, adenine, thymine.

5. State the one gene–one enzyme hypothesis and indicate how it has been modified.

6 Give three differences between DNA and RNA.

7 Name the three types of RNA, and indicate where each is synthesized and where each is active.

8 Discuss the function of each of the following in protein synthesis: DNA, mRNA, rRNA, tRNA, ribosomes, amino acids.

9 Describe the processes of transcription and translation, and explain how genes control cellular functioning.

10 Write the amino acid sequence of a tripeptide using the genetic code on page 644 of your text and given the following nitrogenous base sequence in the DNA molecule: TACGCATCC.

11 Define mutation, give four examples of different types of mutations, and indicate how mutations can be induced.

12 Discuss various possible definitions of "gene" and give advantages and disadvantages of each.

13 Define extrachromosomal inheritance and explain how it occurs in bacteria.

14 Define and use correctly the following terms: virulent phage, temperate phage, episome, plasmid, provirus.

15 Discuss the lytic and lysogenic cycles of bacteriophage viruses.

16 Explain how plasmids are used in recombinant DNA technology, and cite the advantages and possible dangers of this technology.

SUMMARY

Experiments on bacterial transformation and radioactively labeled bacteriophage demonstrated that DNA, not proteins, constitutes the genetic material. DNA is composed of building blocks called *nucleotides*, which are made up of a five-carbon sugar attached by covalent bonding to a phosphate group and a nitrogenous base:

There are four different nucleotides in DNA; they differ in their nitrogenous bases, which may be the double-ring purines, *adenine* and *guanine*, or the single-ring pyrimidines, *cytosine* and *thymine*.

The structure of DNA and its replication Using information gained from chemical analysis, physical chemistry, and X-ray diffraction studies, James Watson and Francis Crick formulated a model of the DNA molecule. According to this model, the nucleotides are joined together by covalent bonds between the sugar of one nucleotide and the phosphate group of the next nucleotide in the sequence; the nitrogenous bases are side groups of the chains. DNA molecules are usually double-chained structures, with the two chains held together by hydrogen bonds between adenine and thymine from opposite chains and between guanine and cytosine from opposite chains. The ladderlike double-chained molecule is coiled into a double helix.

The Watson-Crick model of DNA explains how genetic replication can occur. The two chains of the DNA molecule are separated in places, and each chain acts as a mold or template for the synthesis of its new partner. The process produces two complete double-chained molecules, each identical in base sequence to the original double-chained molecule.

The mechanism of gene action DNA also controls cellular function. Using evidence from the mold *Neurospora*, George W. Beadle and Edward L. Tatum proposed the one gene–one enzyme hypothesis, which states that each gene directs the synthesis of a protein enzyme that controls a chemical reaction of the cell; the reactions, in turn, determine the phenotypic characteristics. Because some proteins are composed of two or more chemically different polypeptide chains, each determined by its own gene, the hypothesis has been modified to a new form: one gene–one polypeptide.

The sequence of bases in DNA indicates the sequence in which amino acids must be linked in protein synthesis. When the double-stranded DNA that constitutes a particular gene is activated, its two nucleotide chains uncoil partially, and one of the chains acts as the template for the synthesis of single-stranded *messenger RNA* (mRNA). This process is called *transcription*; the information encoded in the nucleic acid of the DNA gene is transcribed into mRNA. This mRNA now leaves the nucleus and moves into the cytoplasm, where it becomes associated with the ribosomes.

The mRNA carries information for protein synthesis, but the nucleic acid message must be translated into an amino acid sequence. The coding unit or *codon* in the nucleic acids is three nucleotides long. All but three of the 64 possible triplet codons code for one of the 20 amino acids. The genetic code is degenerate and nonoverlapping.

Protein synthesis The various types of *transfer RNA* (tRNA) in the cytoplasm pick up their specific amino acids and bring them to a ribosome as it moves along the mRNA. Each tRNA attaches to the mRNA at the point where a triplet of mRNA bases (a codon) is complementary to the exposed triplet (*anticodon*) on the tRNA. This ordering of the tRNAs along the mRNA automatically orders the amino acids, which are then linked by peptide bonds. This process is called *translation*; the nucleic acid message has been translated into an amino acid sequence.

Synthesis of the polypeptide chain proceeds one amino acid at a time in an orderly sequence as the ribosomes move along the mRNA. As each tRNA donates its amino acid to the growing polypeptide chain, it uncouples from the mRNA and moves away to pick up another load. When a ribosome reaches a termination codon, it releases the completed polypeptide chain.

In this process, the DNA of the gene determines the mRNA, which determines the protein enzymes, which control chemical reactions, which produce the characteristics of the organism.

Mutations Mutations are alterations in the DNA that change its information content and thus produce new alleles. Several types of mutations are possible. The *addition* or *deletion* of nucleotides in DNA often results in the production of inactive enzymes because of frame shifts in the translation process. In *base substitution* (point mutation), one nucleotide is exchanged for another. Base-substitution mutations are often due to mutagenic agents but may occur spontaneously as a result of mispairing of bases during DNA replication. Such *intragenic recombination* may also occur as a result of crossingover at points within the gene.

High-energy radiations and a variety of chemicals can cause genetic mutations. Ionizing radiations sometimes induce point mutations, but frequently produce large deletions of genetic material. Some mutagenic chemicals convert one base into another. There is a strong relationship between the mutagenicity of a chemical and its cancer-inducing activity (carcinogenicity).

Definitions of the gene A gene can be defined in many ways. It may be defined as a location on the chromosome that exerts control over one or more characteristics of an organism. Or it may be considered as the smallest unit of recombination and mutation, which we now know would make the gene only one nucleotide long, far smaller than envisioned in the classical definition based on recombination. Biochemical geneticists usually define the gene as a unit of function, or *cistron*, which is the length of DNA that determines one polypeptide chain. Since some genes act as templates for tRNA and rRNA synthesis, it would be more accurate to define the gene as the length of DNA that codes for one functional product. However, much of the chromosomal DNA is nongenic, and its function is unknown.

Extrachromosomal inheritance Such cytoplasmic structures as plastids, mitochondria, centrioles, and the basal bodies of cilia and flagella have their own DNA and replicate themselves. Their characteristics are at least partly determined by their own genes.

Bacterial cells often contain, in addition to their main circular chromosome, other small circular pieces of DNA called *plasmids*. These are free in the cytoplasm and replicate independently.

Bacteria are able to exchange genetic material through the process of *conjugation*. Two bacterial cells lie very close to one another, and a cytoplasmic bridge forms between them. Genetic material can pass through this bridge. Conjugation can occur only between cells of different mating types. The cells of one mating type contain a sex factor, a type of plasmid. When the sex factor is free in the cytoplasm, it can easily be transferred from cell to cell by conjugation. At times, however, the sex factor is inserted into the circular bacterial chromosome. When conjugation occurs, synthesis of a linear chromosome on the template of the circular chromosome begins and the new linear chromosome moves through the conjugation bridge into the recipient cell. Conjugation usually stops before the entire chromosome is transferred. Since the chromosome moves at a fairly steady rate, disrupting conjugation at measured intervals permits mapping of the chromosomal genes. The sex factor is sometimes free in the cytoplasm; at other times it is integrated into the bacterial chromosome. Plasmids that exhibit this dual behavior are called *episomes*.

Bacteriophage viruses may also act as episomes. The DNA of virulent phages in the cytoplasm takes control of the cell's metabolic machinery and puts it to work synthesizing new viruses. The bacterial cell soon lyses (bursts), hundreds of new viruses are released to attack other cells, and the *lytic cycle* is repeated. However, the DNA of some viruses can be present in the cell in an inactive state; integrated into the bacterial chromosome, it functions and replicates as an additional part of the bacterial chromosome (*lysogenic cycle*). The virus in this state is in the *provirus* form. Under unfavorable conditions the provirus may leave the chromosome and begin the lytic cycle. Sometimes bits of the bacterial chromosome become incorporated into new infectious viruses as the virus is replicated. Such viruses will inject both viral and bacterial DNA into the cells they infect, and the bacterial genes can undergo recombination with the host's genes. The virus has thus carried genes from one bacterial cell to another—a process called *transduction*. Viral DNA can also act as an episome in the cells of higher organisms.

Most bacterial cells also contain plasmids that cannot integrate into the bacterial chromosome; they remain autonomous. Many such plasmids bear genes for antibiotic resistance. Plasmids can be used to transfer genes from one kind of organism into the cell of another, using a technique called *recombinant DNA technology*. New genetic material can be added to plasmids, and these modified plasmids can be picked up by bacterial cells, which thereby acquire the foreign genes.

QUESTIONS

Testing recall

Below are listed some of the important events that led to our present knowledge of the nature of the gene and its action. Match each of these with the investigator(s) associated with it.

a Beadle and Tatum
b Chargaff
c Griffith
d Hershey and Chase
e Meselson and Stahl
f Leder and Nirenberg
g Watson and Crick
h Wilkins

1 proposal of the double-helix model of DNA

2 transformation of bacteria by material extracted from heat-killed virulent cells

3 infection of bacteria using radioactively labeled bacteriophage

4 discovery that in any DNA the amount of A equals the amount of T and the amount of G equals the amount of C

5 X-ray diffraction studies of DNA

6 construction of the genetic dictionary

7 enunciation of the one gene–one enzyme hypothesis

8 use of N^{15}-labeled DNA to obtain evidence in support of the Watson-Crick mechanism of DNA replication

Match each item below with the appropriate term. A term may be used more than once or not at all.

a sugar-phosphate groups
b purine(s)
c pyrimidine(s)
d covalent bonds
e hydrogen bonds
f mRNA
g tRNA
h rRNA

9 backbone of the DNA molecule

10 force between the two polynucleotide chains

11 single-ring nitrogenous bases

12 double-ring nitrogenous bases

13 adenine and guanine

14 cytosine and thymine

15 uracil

16 formed as a result of transcription

17 carrier of amino acids to the ribosome

18 carrier of the genetic message to the ribosome

19 One strand of a DNA molecule has the sequence of bases TACCTTCAGCGT.

   a What is the sequence of bases on the complementary strand of DNA?
   b What is the sequence of bases on the strand of mRNA that is synthesized from the original strand?
   c Name the organelle where the synthesis of mRNA takes place.
   d How many codons are there in this strand?
   e What are the anticodons for the mRNA transcribed from this segment of DNA?
   f Name the organelle where the codon and anticodon couplings take place.
   g Given the following information

Anticodon	Amino acid
CUU	glutamic acid
UUC	lysine
UAC	methionine
AGC	serine
CGU	alanine
CAG	valine

   what is the sequence of the amino acids in the polypeptide formed from this DNA strand?

Testing knowledge and understanding

Choose the one best answer.

1 The structure of DNA as proposed by Watson and Crick *depended* on all of the following observations *except*

   a that DNA is capable of replicating itself precisely.
   b that DNA base sequences vary from organism to organism.
   c that DNA contains nitrogenous bases, sugars, and phosphates.
   d X-ray periodicities of 3.4 nm, 2 nm, and 0.34 nm.
   e Chargaff's rules of A=T and G=C.

2 Which of the following is *not* compatible with the concept that DNA is the genetic material?

   a DNA content of nuclei is constant in the cells of any one species but is only half the usual amount in the nuclei of the gametes.
   b Each species has equal amounts of adenine, thymine, guanine, and cytosine.
   c DNA content of the nuclei doubles before division.
   d Only the DNA of a bacteriophage enters a new host bacterium.

3 A nucleic acid is composed of a chain of nucleotides. Nucleotides themselves are made of three components. Which of these components could be removed from a nucleotide that is part of a nucleic acid chain without breaking the chain?

   a sugar
   b phosphate
   c nitrogenous base

4 A parent molecule of DNA containing only radioactive nitrogen N^{15} is placed in an environment containing only N^{14}. After four replications, how many DNA molecules would still contain some N^{15}? (Assume no crossing-over.)

   a 2
   b 4
   c 6
   d 8
   e 16

5 DNA, but not RNA, contains

   a adenine.
   b guanine.
   c thymine.
   d cytosine.
   e uracil.

6 According to current ideas about the DNA genetic code, which one of the following statements is *not* true?

   a The codon is three nucleotides long.
   b Every possible triplet codes for some amino acid.
   c The code is redundant (i.e. it contains "synonyms").
   d The code is read in a regular sequence, beginning at one end.
   e The code is nonoverlapping.

7 If a segment of nucleic acid is CATCATTAC, the complementary DNA strand is

   a CATTACTAC.
   b CAUCAUUAC.
   c GUAGUAAUG.
   d GTAGTAATG.
   e CUACUACAT.

8 If mRNA is transcribed from a strip of DNA with the base sequence CATTAG, the mRNA will have the sequence

a GUAAUC.
b GTAATC.
c TGCCGA.
d TUCCUA.
e GTUUTC.

9 Transfer RNA functions in

a carrying RNA from the ribosomes to mRNA.
b attaching RNA to the ribosomes.
c joining proteins to form the ribosomes.
d carrying mRNA from the nucleus to the cytoplasm.
e carrying amino acids to the correct site on the mRNA.

10 If tRNA specialized for transfer of the amino acid methionine has the anticodon UAC, what codon on the mRNA codes for methionine?

a UAC
b TAC
c AUG
d ATG
e ACU

11 In the situation above, the DNA codon for methionine must be

a TAC.
b AUG.
c UAC.
d ATG.
e TCA.

12 A certain gene codes for a polypeptide that is 126 amino acids long. The gene is probably how many nucleotides long?

a 42
b 126
c 252
d 378
e 504

13 Suppose a gene has the DNA nucleotide sequence

1 2 3 4 5 6 7 8 9 10 11 12 13 14 15 16 17
C T G G C A T G C T T C G G A A A

(No real gene could be this short, but for our purposes this will suffice.) Which one of the following mutations would probably produce the greatest change in the activity of the protein for which this gene codes?

a substitution of A for G in position 3
b deletion of the C at position 5
c deletion of the A at position 16
d addition of a G between position 14 and 15

14 Which one of the following is *not* a form of recombination?

a translocation
b transformation
c base substitution
d addition mutation
e transduction

15 Although a gene codes ultimately for all aspects of a protein's structure, it codes *directly* only for

a primary structure.
b secondary structure.
c tertiary structure.
d quaternary structure.

16 The direction of transfer of genetic information in most living things is

a protein ⟶ DNA ⟶ mRNA.
b DNA ⟶ mRNA ⟶ protein.
c DNA ⟶ tRNA ⟶ protein.
d protein ⟶ tRNA ⟶ DNA.
e RNA ⟶ DNA ⟶ mRNA ⟶ protein.

17 Which one of the following statements is *not* true?

a The nucleolus is a specialized region of a chromosome where rRNA is synthesized.
b In point mutations only a single nucleotide of a gene is altered.
c Molecules of mRNA are synthesized on the ribosomes from nucleotides brought by tRNA.
d Some amino acids are specified by several "synonymous" codons.

18 Bacteria sometimes demonstrate a type of genetic recombination in which a virus acts as an agent transporting DNA from one bacterial cell to another. This process is called

a conjugation.
b transformation.
c transduction.
d translocation.
e inversion.

19 Which one of the following statements is *not* true?

a Live bacteria can sometimes pick up DNA that has been released into the medium from dead cells; this process is called transformation.
b Sometimes two bacterial cells come to lie very close to each other and form a cytoplasmic bridge; the DNA moves from one cell into the other.
c Sometimes when new infectious virus particles are formed in a host cell, some host DNA may be enclosed within one of the particles; when the virus attacks a

new host, the DNA from the first host may be injected into the second host in the process called transduction.

d Viral DNA that has been incorporated as provirus into the chromosome of a host cell can remain there only until the cell divides, because proviral DNA cannot be replicated prior to mitosis of the host cell.

e When new infectious viral particles are manufactured in a bacterial cell, they are eventually released into the surrounding medium when the cell undergoes lysis.

For further thought

1 A nonsense mutation is one in which the substitution of one nucleotide for another results in a triplet that does not code for any amino acid. A missense mutation is one in which such a substitution results in a triplet that codes for a different amino acid. What effect would each of these have on protein synthesis? Which of these would be expected to have the most severe effect on enzyme activity?

2 Active beef insulin is a small protein composed of 51 amino acids. How many nucleotide units does it take to code for this protein? Using the diagram of beef insulin on page 58 of your text and the genetic code on page 644, determine the nucleotide sequence in the DNA that codes for the shorter chain.

3 Describe as many types of genetic recombination as you can. Which of these are known to occur in human beings?

Chapter 16

DEVELOPMENT: ASPECTS OF CELLULAR CONTROL

KEY CONCEPTS

1 Since all the cells of a multicellular organism are normally genetically identical, control mechanisms determine which of any cell's inherited instructions will be acted upon and which will not.

2 Each gene codes for only one kind of messenger RNA; regulators determine if and when each gene will synthesize its particular messenger RNA.

3 Environmental influences can act by helping turn on or off the synthetic activity of the various genes or by influencing protein synthesis and activity.

4 Cancer cells proliferate without restraint; the normal cellular control mechanisms no longer work.

5 An important defense against disease in vertebrate animals is their ability to manufacture antibodies that can inactivate or destroy foreign substances.

OBJECTIVES

After studying this chapter and reflecting on it, you should be able to

1 Explain why cells from different parts of an organism have different structural and functional characteristics even though they are genetically identical.

2 Describe the Jacob-Monod operon model for substrate induction; include in your description the role of the inducer, operator, promoter, repressor protein, regulator gene, and structural genes.

3 Differentiate between inducible and constitutive enzymes.

4 Explain how end-product corepression and the cAMP-CAP complex control gene transcription, and how these mechanisms differ from control by substrate induction.

5 Give a reason why the mechanisms for the control of gene transcription found in bacteria are not directly applicable to the eucaryotic cell.

6 Name two substances that may act as selective agents of gene control in eucaryotic cells.

7 Describe the structure of the giant salivary chromosomes of *Drosophila* and explain why some areas of the chromosome puff out.

8 Define gene amplification and relate this process to the formation of the nuclear and extrachromosomal nucleoli.

9 Describe the points at which cellular control can be applied in the path of information flow from the genes to their phenotypic expression.

10 Give three principal differences between cancer cells and normal cells.

11 Compare the action of a cancer-inducing virus in a susceptible and nonsusceptible cell.

12 Explain what happens when an organism is exposed to an antigen and is stimulated to make antibodies.

13 Differentiate between B lymphocytes and T lymphocytes.

14 Describe the hypothesis of clonal selection.

15 Explain why the first exposure to an antigen produces a delayed response whereas subsequent exposures produce an accelerated response.

16 Draw a diagram of the antibody molecule and label light chains, heavy chains, constant and variable ends, and the antigen binding site.

17 Differentiate between the germline hypothesis and the somatic mutation hypothesis.

18 Give a hypothesis to account for the way the body recognizes "self" and "not self." Explain how this process relates to immunologic diseases.

SUMMARY

Although every cell in the body of a multicellular organism ordinarily has identical genetic information, each has different structural and functional characteristics. The activities of any single cell also differ at different times. Various control mechanisms determine when and how each cell will act on its inherited genetic instructions.

The control of gene transcription in bacteria Cells may select which of their many inherited instructions they use. Bacterial cells have many systems that control gene transcription; three of the best understood are substrate induction, end-product corepression, and cAMP-CAP activation.

The Jacob-Monod model of *substrate induction* in bacteria proposes that three parts of the chromosome are involved in controlling transcription of the *structural genes*: the repressor gene and the operator and promoter regions. According to this model, transcription of the structural genes occurs only when the substrate is present: the substrate acts as an *inducer* and turns on a particular *operon* by inactivating the *repressor* protein, thereby freeing the *operator* of repression. Now RNA polymerase can bind to the *promoter* and initiate structural gene transcription. The mRNA produced moves to the ribosomes, translation occurs, and enzymes are synthesized.

In *repressible enzyme* systems the operons are always on and enzymes are automatically synthesized unless the operator is turned off by repressor protein activated by *corepressor substances* (often the end-product of a biochemical pathway).

Control of gene transcription by the *cAMP-CAP complex* is a positive mechanism. When the cAMP level increases in the cells, it binds to CAP (the activator protein). The cAMP-CAP complex then binds to and activates the promoter region, stimulating transcription.

Control of gene transcription in eucaryotes The control of gene transcription in eucaryotic cells differs from that in bacteria because the DNA of eucaryotic chromosomes is tightly complexed with proteins to form *chromatin*. These chromosomal proteins are active in gene regulation.

The *histones* appear to be structural elements of the chromosomes and may play a nonselective role in the coiling and uncoiling of the chromosomes and in masking DNA. Evidence indicates that nonhistone *acidic proteins* or RNA may act as selective agents in gene regulation. Their activity is modulated by chemicals from the cytoplasm.

When stained, eucaryotic chromosomes can be differentiated into *euchromatic* and *heterochromatic* regions. The euchromatic regions contain active genes; heterochromatic regions are inactive. Even though eucaryotic genes are not organized as operons, the synthesis of some groups of enzymes can be controlled as a unit. In these cases, the original polypeptide is very long and is cut into several enzymes. Other mechanisms for coordinating induction and repression are unknown.

Lampbrush chromosomes and *chromosomal puffs* are visible evidences of gene activity. The DNA of the active regions of the chromosomes loops out laterally from the main chromosome. The pattern of chromosomal puffs varies with the developmental stage and with changes in the extranuclear environment, showing clearly that changes in the cytoplasm can alter gene activity.

Experiments indicate that many environmental influences act by turning on or off the synthetic activity of the various genes. Each gene can make only one mRNA; regulators determine if and when each gene will synthesize its mRNA.

Gene amplification is another pre-transcriptional control mechanism. In this process multiple copies of the genes for rRNA are made, in addition to the many copies of those same genes that already compose the nucleolus.

Post-transcriptional control A variety of post-transcriptional processes may act as points of cellular control. For example, transcribed eucaryotic mRNA apparently does not operate until enzymes in the nucleus convert pre-mRNA molecules into functional mRNA. This conversion may be a point of control. Control may also be exerted through the rate at which mRNA is broken down within the cell.

Protein synthesis and the regulation of protein activity provide additional points for cellular control. Control may also be exerted by regulating the rate of conversion of some proteins into their functional form, by influencing the self-assembly of some proteins, or by regulating the rate of enzyme destruction. The physical and chemical environment also affects cellular activity by influencing enzyme catalysis.

Cancer: A failure of normal cellular controls The single most distinctive feature of cancer cells is their unrestrained growth and division. Cancer cells are almost always genetically unlike normal cells in that they have extra chromosomes. When nutritionally limited, cancer cells lack the control mechanism to stabilize in the G_1 stage of the cell cycle as normal cells do. Instead the individual cells stop randomly in any of the stages of the cell cycle, which makes them more susceptible to stressful factors.

Cancer cells also show many differences related to cell surfaces. They are more spherical in shape and more mobile than normal cells. Their cell surface differs from that of a normal cell. Because of their mobility, they do not show contact inhibition of cell growth and movement. Nor can cancer cells recognize other cells of their own type. The surfaces of cancer cells characteristically bear antigens not found on normal cells.

Evidence indicates that some cancers are caused by viruses. When a cancer virus invades a cell of its normal host species, it usually causes the cell to make new virus particles. But when the virus invades the cell of another species incapable of making that virus, it may become integrated as a provirus and *transform* the cell into a cancerous condition. Virus-induced cancers may result from a mismatch between a virus and the host cell it has invaded.

The immune response Vertebrate animals can manufacture highly specific *antibodies* to inactivate or destroy invading *antigens*, which are large molecules ordinarily foreign to the organism's body. The cells that respond to the foreign antigen are the *lymphocytes*. Two types of lymphocytes can be recognized: *B lymphocytes*, which are derived from stem cells in the bone marrow, and *T lymphocytes*, which are derived from stem cells in the thymus. When stimulated, the B cells produce circulating antibodies, resulting in *humoral immunity*; the antibodies of the T cells remain bound to the cell and the whole antibody-bearing cell attacks the antigen (*cell-mediated immunity*). The functions of the B cells and T cells are not entirely separate;

many T cells appear to play a regulatory role in the responses of B cells to antigens.

Only certain regions of the large antigen molecules serve as *antigenic determinants*, the sites of interaction with the lymphocyte receptors. Each lymphocyte, whether of the B or T type, has one type of exposed antibody on its surface. An antigen can bind only to those lymphocytes whose surface antibodies are specific for its antigenic determinant regions. According to the hypothesis of *clonal selection*, each stimulated lymphocyte begins to enlarge and proliferate, producing a clone. In the clones are antibody-producing plasma cells if the stimulated cells were B lymphocytes, or immunologically active T cells if the stimulated cells were T lymphocytes. Also in the clones are many *memory cells* that may persist for the organism's lifetime, conferring a lasting immunity to the stimulating antigen.

Antibodies are globulin proteins. Each molecule consists of four polypeptide chains: two identical "heavy" chains and two identical shorter "light chains," linked by disulfide bonds. Most of the antibody molecule is constant in its amino acid composition; the two binding sites for the antigens are at the ends of the variable portion of the molecule.

Two principal hypotheses have been advanced to account for the genetic basis of antibody diversity. The *germline hypothesis* states that every lymphocyte carries genes for all possible antibodies, and that in each lymphocyte only one antibody gene is active; the others are repressed. The *somatic mutation hypothesis* states that each cell has only a few antibody genes, but that certain regions on these genes are extremely susceptible to mutation. Thus each lymphocyte carries its own unique mutant form.

We are not sure how the body distinguishes between "self" and "not self." It appears that the mature organism recognizes as "self" those substances that were present at the critical time in the organism's development; substances that were not present at the critical time are "not self." However, the capacity for developing self-tolerance continues to some degree throughout life.

Occasionally the immunologic mechanisms get out of control, and the ability to distinguish between "self" and "not self" is impaired. In such cases the body begins to destroy itself (auto-immune diseases).

QUESTIONS

Testing recall

Which of the following are associated with increased gene transcription?

1 binding of the inducer to the repressor

2 formation of the cAMP-CAP complex

3 binding of the corepressor to the repressor protein

4 binding of the RNA polymerase to the promoter

5 enzymatic conversion of pre-mRNA

6 gene amplification

7 puffing of the chromosomes

8 end-product corepression

9 formation of lampbrush chromosomes

Mark each of the following statements T if it is generally associated with the T lymphocytes and B if it is generally associated with the B lymphocytes. Mark B and T if it is associated with both types of lymphocyte.

10 cellular immunity

11 humoral immunity

12 plasma cells

13 circulating antibodies

14 cell-bound antibodies

15 memory cells

16 formed from stem cells in the thymus

17 formed from stem cells in the bone marrow

18 generally stimulated by viruses and molecular antigens

Indicate whether each of the following is more characteristic of cancer cells (C) or normal cells (N).

19 extra chromosomes

20 stabilization in the G_1 stage when nutritionally deprived

21 unrestrained proliferation

22 abnormal cell surface

23 recognition of cells of their own tissue type

24 spherical shape

25 contact inhibition of growth and cell division

26 excess glucose consumption with increased lactic acid production

27 secretion of large amounts of proteolytic enzymes

Testing knowledge and understanding

Choose the one best answer.

1 According to the Jacob-Monod (lac operon) model of gene regulation, inducer substances in bacterial cells probably

   a combine with operator regions, activating the associated operons.
   b combine with structural genes, stimulating them to synthesize messenger RNA.
   c combine with repressor proteins, inactivating them.
   d combine with promoter regions, activating RNA polymerase.
   e combine with nucleoli, triggering production of more ribosomes.

2 The promoter region of a bacterial operon

   a codes for repressor proteins.
   b codes for inducer substances.
   c codes for corepressor substances.
   d is a binding site for inducers.
   e is a binding site for RNA polymerase.

3 In bacteria, the structural genes can be turned off when

   a the end-product of a reaction combines with the repressor protein and activates it.
   b the substrate combines with a repressor protein and inactivates it.
   c the substrate combines with the repressor protein and activates the promoter.
   d the substrate combines with its enzyme and produces a repressor molecule.
   e the cAMP-CAP complex binds to the promoter.

4 If an insect is given the hormone ecdysone, certain regions of the insect's chromosomes will soon exhibit puffing. What important process is taking place in the puffs?

   a mRNA synthesis   d enzyme synthesis
   b rRNA synthesis   e histone synthesis
   c DNA replication

5 Which one of the following statements about immunity is false?

   a The vertebrate body apparently has all the genetic information necessary to make antibodies before it encounters stimulating antigens.
   b Each individual B lymphocyte can make several different kinds of antibodies.
   c The T lymphocytes function primarily in cell-mediated immune responses.
   d When an antigen enters a host organism, it combines with cells bearing surface antibodies complementary to it, and initiates proliferation of these cells.
   e The plasma cells formed during an immune response produce the circulating antibodies.

6 Which organ or tissue is *not* an integral part of the immune system?

   a pancreas   d spleen
   b lymph nodes   e bone marrow
   c thymus

7 Which statement about antibodies is *false*?

   a Each antibody combines with a specific antigen.
   b Two sites on each antibody molecule can bind antigen.
   c Antibodies are globulin proteins.
   d Antibodies are found in vertebrate animals.
   e Human blood contains high levels of circulating antibodies against all possible antigens.

8 The memory cells of the immune system

   a are nonspecific; each reacts to a variety of antigens.
   b produce circulating antibodies.
   c are produced by the thymus.
   d are responsible for an accelerated response upon second exposure to an antigen.
   e remain in circulation for a short time and are then destroyed.

For further thought

1 Cancer cells seem to lack the normal control mechanisms for growth and cell division. Scientists hope that by studying the control of normal cell growth and division they will gain some understanding of why these controls are not exercised in cancer cells. At which points in the flow of information from gene to phenotype do normal controls operate? At what points do these controls appear to be lacking in cancer cells?

2 Radiation is extensively used to treat cancerous tumors. Explain why radiation can destroy cancer cells while allowing normal cells to continue to function.

3 In what ways do aberrations of the cell surface contribute to the properties of cancer cells?

4 The Sabin poliomyelitis vaccine consists of three strains of weakened live viruses.

   a Explain the process by which your body develops immunity to the polio viruses after you receive an oral dose of this vaccine.

   b Why are babies usually given two or three separate doses of the vaccine rather than one dose?

   c Children are given a "booster" vaccine at the age of 1½ years and again before school. What is the purpose of the boosters?

5 It has been suggested that a number of human degenerative diseases, such as rheumatoid arthritis and multiple sclerosis, may be the result of a slow-virus infection. In a slow-virus infection the virus can remain in the cells and replicate, while allowing the infected cell to continue to function. However, the slow virus does alter the protein composition of the cell surface. Explain how the altered cell surface may be responsible for much of the damage found in degenerative diseases.

Chapter 17

DEVELOPMENT: FROM EGG TO ORGANISM

KEY CONCEPTS

1 Development in sexually reproducing organisms begins with the penetration of the sperm into the ovum and ends with the organism's death.

2 **Development proceeds step by step and is** generally irreversible.

3 In animals, the developmental processes of cell division, cell growth, cell differentiation, and morphogenetic movements convert the fertilized egg into the mature organism.

4 In plants, the same developmental processes occur except for morphogenetic movements; in plants morphogenesis occurs by the patterns of cell division and growth.

5 In plants a few cells remain forever embryonic; thus new organs can be formed and growth will continue throughout the plant's life.

6 Because all the somatic cells of a single organism arose by repeated division of the fertilized egg and are genetically identical, substances in the cytoplasm and the cellular environment determine which potentialities are expressed.

7 The cytoplasmic substances to which a nucleus is exposed play a prominent role in activating some genes and repressing others; therefore nonuniform distribution of cytoplasmic substances in dividing cells can restrict the development of cells and their progeny.

8 As cells and tissues become more differentiated, they alter the environment of other cells near them through the chemicals they secrete; these changes in the cellular environment profoundly affect gene activity.

9 Differentiation is a matter of progressive determination; development is gradually restricted to one of the many initially possible pathways.

OBJECTIVES

After studying this chapter and reflecting on it, you should be able to

1. Describe the events triggered by the penetration of an animal egg by the sperm, and discuss the process of fertilization.

2. Describe the principal events that occur during the early cleavage stages of an animal embryo and explain why these stages are largely controlled by the maternal genes.

3. Define these terms and use them correctly: zygote, cleavage, morula, blastula, blastocoel, gastrula, archenteron, blastopore.

4. Describe gastrulation in amphioxus, and indicate how the process differs in frogs and in birds. Indicate the ultimate fate of the three germ layers.

5. Describe the role of growth in the postembryonic development of an animal.

6. Define metamorphosis and give the adaptive significance of the larval stage. Distinguish between complete and gradual metamorphosis in insects.

7. Give four factors that contribute to the aging of irreplaceable tissues.

8. Discuss three explanations of why aging occurs.

9. Describe the early embryology of an angiosperm plant.

10. Draw and label a diagram of a bean seed.

11. Give the function of each of the following: epicotyl, hypocotyl, cotyledon, endosperm, radicle, seed coat.

12. State four major differences between embryonic development in an angiosperm plant and an animal.

13. Describe the processes involved in the growth of the root and stem of a young plant.

14. Briefly explain the role auxin is thought to play in inducing cell wall plasticity and in insertion of new wall material.

15. Describe the process by which leaves are produced along the growing young stem.

16. Sketch the tip of a growing root or stem, showing the areas of cell division, cell enlargement, and cell maturation.

17. Contrast the function of the apical meristems with that of the lateral meristems.

18. Explain, using an example, how the polarity of the egg cell and the plane of cleavage can influence development.

19. Distinguish between determinate and indeterminate cleavage, and name animals that show each pattern.

20. List three stimuli that are known to influence the patterning of the egg cytoplasm.

21. Describe embryonic induction, using the dorsal lip of the blastopore in the frog as an example.

22. Distinguish between instructive and permissive inducers, and classify plant and animal hormones according to their inductive role.

23. Explain the concept of morphogenetic fields.

24. Cite evidence to indicate that cells have positional information to ensure proper pattern formation.

25. Discuss the role of the physical environment in guiding development.

26. Define differentiation, and use C. H. Waddington's developmental landscape model to explain the gradualness of differentiation. Give experimental evidence that supports this model.

27. Cite experimental evidence from plants and animals that shows whether differentiation is reversible or irreversible.

SUMMARY

Development of a multicellular animal The process of development in a sexually reproducing multicellular animal begins with the penetration of the sperm into the ovum. The membrane of the sperm fuses with that of the egg and the sperm nucleus moves into the egg. Vesicles in the outer region of the egg then release their contents, forming the *fertilization membrane*, which prevents penetration by other sperm. Penetration also initiates the completion of oogenic meiosis. *Fertilization* occurs when the two gamete nuclei fuse.

The zygote then undergoes a rapid series of mitotic divisions called *cleavage*. These early cleavages are not accompanied by protoplasmic growth; the cytoplasm of the one large cell is partitioned into many new smaller cells (*blastomeres*). As cleavage continues, the cells of chordates become arranged in a hollow ball called a *blastula*. Two regions are often evident: the *animal hemisphere*, made up of small dark cells, and the *vegetal hemisphere*, made up of larger yolky cells.

Next begins a series of complex movements that are important in establishing the shape and pattern of the developing organism (*morphogenesis*). The blastula is converted into an embryo called the *gastrula*. Cleavage and gastrulation are greatly influenced by the amount of yolk in the egg. Gastrulation first produces an embryo with two layers, an outer *ectoderm* and an inner *endoderm*. A third layer, the *mesoderm*, forms between them. The ectoderm gives rise to the outermost layers of the body, the nervous system, and the sense organs; the endoderm to the lining of the digestive tract and associated structures; and the mesoderm to the supportive tissue—muscles and connective tissues.

The morphogenetic movements of gastrulation and neurulation give form and shape to the embryo, and bring masses of cells into proper position for their later differentiation into the principal tissues of the adult body. The developmental processes of cell division, cell growth, cell differentiation, and morphogenetic movement convert the gastrula into a young animal ready for birth.

The predominant factor in postembryonic development in most animals is growth in size. Growth does not occur at the same rate and at the same time in all parts of the body.

Many aquatic animals and certain groups of terrestrial insects go through *larval* stages that bear little resemblance to the adult. The complex series of developmental stages that convert an immature animal into an adult is called *metamorphosis*. Some insects undergo *complete metamorphosis*, which begins with a wormlike larval stage quite unlike the adult. After a period of growth the larva enters the *pupal stage*, during which it is extensively reorganized to form the adult. Other insects undergo *gradual metamorphosis*; the young, which already resemble the adult, go through a series of molts that make them more and more like the adult.

Aging is another aspect of development; it is a complex series of developmental changes that lead to the deterioration of the mature organism and ultimately to its death. In both plants and animals the aging process seems to be correlated with the degree of cellular specialization; cells that remain relatively unspecialized and continue to divide do not age as rapidly as cells that have lost the capacity to divide. Aging of the whole multicellular organism results from the deterioration and death of irreplaceable cells and tissues.

Aging in animals seems to be related to such factors as the replacement of diseased or injured tissue by connective tissue, the increased burden on remaining cells, altered hormonal balance, the deposition of waste material in older cells, and changes in the intercellular connective tissue. Scientists cannot yet explain fully why these changes occur.

Development of an angiosperm plant The egg cell of an angiosperm plant is fertilized in the ovary of the maternal plant by a sperm nucleus from a pollen grain. The first cell division gives rise to two unequal cells, a terminal cell and a basal cell. The terminal cell will give rise to the embryo proper. Three types of tissues begin to differentiate: a surface layer of *protoderm*, which will form epidermal tissue; an inner core of *provascular* tissue, which will form the cambium and vascular tissues; and a middle layer of *ground tissue*, which will form the cortex. Next, two embryo leaves, the *cotyledons*, arise. The part of the embryonic axis below the cotyledons is the *hypocotyl*; it will form the first part of the stem. Small clumps of tissue at each end of the embryonic axis remain undifferentiated and become the *apical meristems* of the shoot and root. The embryo, together with the food-storage tissue, the endosperm, becomes enclosed in a seed coat. The resulting structure is the seed.

Embryonic development of the angiosperm differs from that of most animals in that cell division and cell growth occur together, and morphogenesis is governed by patterns of cell growth and division rather than cell migration. Also, some plant cells remain forever embryonic (meristematic) and organogenesis continues throughout the life of the plant.

When the seed germinates, the hypocotyl emerges and turns downward; the *radicle* at its lower end forms the root. The *epicotyl* then begins to elongate; it forms most of the shoot. Growth of the shoot or root involves the production of new cells by the apical meristem, and then the elongation and differentiation of these cells.

Cell elongation is controlled by hormones. Auxin appears to activate a transmembrane hydrogen pump that transports hydrogen ions from the cytoplasm into the wall, breaking hydrogen bonds between the polysaccharide fibrils. The hydrostatic pressure of the cell contents causes stretching of the loosened walls and the cell elongates, but without synthesis of new cytoplasm. Auxin also seems to influence the insertion of new wall material.

In stems, certain areas of the apical meristem give rise to *nodes*, swellings where leaf primordia arise. The length of stem between two successive nodes is an *internode*. Most increase in stem length results from elongation of the cells in the young internodes. At the tip of the stem is a *bud*, which consists of the apical meristem and unelongated internodes enclosed within leaf primordia. When the bud opens, the internodes elongate and mitosis within the leaf primordia produces the leaves. Small areas of meristematic tissue (the *axillary buds*) arise between the leaf and internode.

The cells produced by the apical meristems will differentiate to form the various primary tissues of the plant. Increase in circumference of a root or stem depends upon the formation of secondary tissues derived from the lateral meristems.

The problem of differentiation and pattern formation The factors that influence which of a given cell's many genetic potentialities it will realize are fundamental in cellular differentiation. Genetic and nongenetic control factors interact to produce differentiated cells, tissues, and organs.

Although the genetic content of all embryonic cells is identical, their cytoplasm is not. The cytoplasm of the zygote is rarely homogeneous. Therefore, the daughter cells produced by cleavage of the egg cell may not share equally in all cytoplasmic materials, depending on the orientation of the first plane of cleavage. If the cleavage partitions critical cytoplasmic constituents (i.e. ones that help activate or repress specific genes) equally, the cleavage is *indeterminate*; the new cells have the full developmental potential of the zygote. These eggs have sometimes been called *regulative eggs*. If, however, the first cleavage partitions critical cytoplasmic constituents unequally, the cleavage is *determinate*; when the daughter cells are separated they do not have equivalent developmental potentialities. Eggs with marked cytoplasmic differences have been called *mosaic* eggs. Indeterminate cleavage is characteristic of most echinoderms and vertebrates; determinate cleavage is characteristic of annelids and molluscs. Even in the species where the first few cleavages are indeterminate, a determinate cleavage eventually occurs.

The mechanisms establishing the polarity of the egg cytoplasm are largely unknown, but in some organisms the point of sperm penetration or various asymmetric environmental conditions are known to play a role. Polarity continues to be an important aspect of organization throughout the life of the organism.

Tissue interactions in development Developing cells are influenced by neighboring cells, by diffusible chemicals they secrete, by hormones, and by many parameters of their physical environment (e.g. temperature, light, humidity, gravity, pressure). Certain *organizer* areas of the embryo (e.g. the *gray crescent* of a frog embryo) have an especially strong inductive function; they control the differentiation of other tissues. Some cell-to-cell interactions may be a direct effect of cell contact, but most developmental interactions are probably chemically mediated. Many chemicals can act as intercellular inducers in embryonic development. Some inducers are instructive and restrict developmental potential; others are permissive and facilitate the already determined potentialities.

In plants, hormones influence nearly all phases of development and may have both instructive and permissive roles. In vertebrates, hormones play a mainly permissive role.

Some chemicals influencing development are inhibitory. As a particular structure develops, it may release substances that inhibit the formation of the same kind of structure in the immediate area (called a *morphogenetic field*). Cells apparently have some sort of positional information available to them that ensures proper pattern formation.

As development proceeds, the individual cells

become more and more committed to one particular course of differentiation. Differentiation is thus a matter of progressive determination; development is gradually restricted to one of the many initially possible pathways.

As cells become more differentiated, the activity of the cells' genetic material changes. In plants, the changes undergone by cell nuclei during differentiation appear to be entirely reversible. In animals, the situation is less clear. Nuclear transplant experiments in frogs indicate that in some cases the differentiated nuclei had apparently lost their former developmental potential; but other experiments produced the opposite result. Somatic cell-fusion experiments also indicate that, at least in some situations, differentiation is reversible.

The regeneration of lost body parts involves dedifferentiation and redifferentiation of cells. The existence of such processes indicates that the nuclei of differentiated animal cells have not undergone irreversible change. Chemical and/or electrical factors may play a role in inducing regeneration.

QUESTIONS

Testing recall

Below are listed some important terms in animal embryology. Match each statement with the proper term.

a animal pole
b archenteron
c blastocoel
d blastopore
e blastula
f gastrula
g ectoderm
h endoderm
i mesoderm
j morphogenesis
k morula
l neurula
m vegetal pole
n zygote

1 grapelike cluster of cells formed by cleavage of the zygote

2 cavity of the blastula

3 opening into the cavity of the digestive tract

4 embryo as a hollow ball

5 process establishing shape and pattern

6 fertilized egg

7 a two-layered, later three-layered, animal embryonic stage

8 cavity of the digestive tract

9 part of the embryo where the invagination of gastrulation usually occurs

10 part that gives rise to the hair, skin, and fingernails

11 part that gives rise to the lining of the digestive tract and pancreas

12 part that gives rise to the brain and spinal cord

13 part that gives rise to muscles and bones

14 Which of the following are correctly matched?

a hypocotyl – radicle
b endosperm – stored food
c morphogenesis in plants – cell movement
d cotyledons – embryonic leaves
e seed dormancy – lowered gibberellin concentration
f epicotyl – primary root
g meristematic cells – differentiated cells
h ground tissue – cortex
i terminal cell – plant embryo
j apical meristem – secondary tissues
k shoot growth – cell division and cell elongation
l auxin – increased plasticity of cell wall
m node – leaf primordia
n vascular cambium – elongation of the stem
o seed – embryo, endosperm, seed coat

15 Which of the following play a direct role in animal cell differentiation?

a pattern of cleavage
b cell migration
c polarity of the egg
d cell division
e induction by chemicals
f hormones
g morphogenetic fields
h physical factors—light, temperature, gravity
i cell elongation

Testing knowledge and understanding

Choose the one best answer.

1 An egg cell differs from a sperm cell because the egg cell

   a has cytoplasm.
   b has mitochondria.
   c contains a haploid set of genes.
   d is a product of meiosis.
   e has much greater energy reserves.

2 The acrosome of the mammalian sperm is

   a derived from mitochondria.
   b derived from Golgi apparatus.
   c the part that contains the nucleus.
   d located in the middle piece.
   e located posterior to the middle piece.

3 The fertilization process can be regarded as completed when

   a a sperm comes into contact with an egg.
   b a fertilization membrane is formed.
   c the sperm penetrates the cell membrane.
   d the sperm enters the egg cytoplasm.
   e the sperm nucleus fuses with the egg nucleus.

4 The process of development begins when

   a the sperm nucleus fuses with the egg nucleus.
   b the sperm penetrates the egg.
   c egg oogenesis is completed.
   d cleavage begins.
   e the cell enters the M stage.

5 In cleavage, with each successive division the

   a number of chromosomes per cell is reduced by half.
   b number of chromosomes is doubled.
   c volume of cytoplasm per cell is reduced.
   d cytoplasmic contents are equally distributed.
   e extensive growth in size of the embryo begins.

6 Which one of the following represents a correct order in the developmental process?

   a zygote ⟶ gastrula ⟶ neurula ⟶ blastula
   b fertilization ⟶ cleavage ⟶ blastula ⟶ gastrula
   c fertilization ⟶ gastrula ⟶ blastula ⟶ neurula
   d zygote ⟶ neurula ⟶ blastocoel ⟶ gastrula
   e fertilization ⟶ zygote ⟶ blastula ⟶ neurula ⟶ gastrula

7 Which one of the following statements is *not* true?

   a In a vertebrate embryo, the morula is formed by mitotic cell divisions without significant cell growth.
   b The embryonic archenteron becomes the lumen of the digestive tract in the adult.
   c Cleavage is restricted to the cytoplasmic disc at the animal pole of the ovum in birds.
   d The details of the gastrulation process vary considerably from species to species, depending on the amount of yolk in the ovum.
   e The embryonic blastocoel opens to the outside via the blastopore.

8 Which one of the following is *not* true of the process of morphogenesis?

   a It refers to the production of form and pattern.
   b It involves cell movements.
   c It plays a major role in the development process.
   d It is synonymous with differentiation.
   e It is exhibited in the formation of the neural tube.

9 As one grows older

   a unspecialized cells age quickly.
   b cell division in all tissues ceases.
   c somatic mutations may accumulate, leading to errors in the DNA.
   d enzymes that repair breaks in DNA become more efficient.

10 Which one of the following processes is important in animal morphogenesis but not in plant morphogenesis?

   a cell division   d cell differentiation
   b cell migration   e cell death
   c cell growth

11 Which one of the following statements about a bean seed and its germination is *not* correct?

   a In the dormant state the seed consumes oxygen, but at a very low rate.
   b As germination begins, much water is absorbed and oxygen consumption rises.
   c The part of the embryonic axis called the epicotyl grows down and gives rise to the plant root.
   d The two cotyledons contain stored food and provide the principal source of nourishment upon which the early growth of the embryo depends.
   e As the young seedling develops, its increase in length is due to mitotic activity in the apical meristems of the root tips and terminal buds, followed by cell elongation in regions immediately behind the meristems.

12 Which one of the following is typical of growth of animals but not of plants?

   a Cell growth is by an increase in the vacuole.
   b Cell growth is by an increase in cytoplasm.
   c Cell division and cell growth typically occur together.
   d Growth continues throughout the life of the organism.
   e Growth is by cell elongation.

13 Which one of the following statements is *not* true?

   a Lateral roots usually arise from the pericycle.
   b Much of the size increase in plant cells involves absorbing more water into the cell vacuole.
   c Provascular tissue is derived from apical meristems.
   d The developed plant embryo contains rudiments of all the organs of the mature plant.
   e Auxin promotes cell elongation.

14 Which one of the following meristematic regions is ordinarily *least* important in increasing the overall size (height and circumference) of a tree over a period of 20 years?

   a apical buds
   b root tips
   c vascular cambium
   d cork cambium

15 Which one of the following areas is closest to the tip of a plant root?

   a axillary bud
   b zone of cell elongation
   c apical meristem
   d zone of cell differentiation
   e zone of cell specialization

16 Which one of the following is *least* likely to be a significant factor in determining the differentiation pattern of embryonic cells?

   a cytoplasmic distribution during early cleavage stages
   b distance from the surface of the embryo
   c inducer substances secreted by neighboring cells
   d the pH and light intensity to which the cells are exposed
   e gene recombination during the cleavage stages

17 Which one of the following inductive influences is inhibitory?

   a morphogenetic fields
   b dorsal lip of the blastopore
   c low concentrations of auxin in plant tissues
   d male sex hormone on the developing male gonad
   e collagen precursor on the presumptive pancreatic tissue

18 Which one of the following statements is *not* true?

   a Visible morphological changes sometimes occur in the regions of a chromosome presumed to be most active.
   b Enucleated frog eggs into which nuclei from adult cells have been transplanted usually (but not always) fail to develop normally.
   c Under some conditions, differentiated plant cells can regain the capacity to divide mitotically.
   d Radioactive tracer studies have shown that many genes are usually lost from the chromosomes of differentiated mammalian cells.
   e Inducer chemicals probably play an important role in determining which genes will be active at any given time.

19 If a *Xenopus* egg is enucleated, and a nucleus is transplanted to it from a more differentiated *Xenopus* cell, what is the most advanced developmental stage from which the nucleus can come if it is to be capable of directing full development of the egg into a normal adult?

   a blastula
   b gastrula
   c neurula
   d tadpole
   e adult

20 The results of transplantation of frog nuclei into eggs allow for which conclusion?

   a The nuclei of differentiated cells generally lack some important genes.
   b As a cell becomes more differentiated, its nucleus is less likely to support development of an egg into a tadpole after nuclear transplantation.
   c The differentiated state is not reversible.
   d The differentiated state is always irreversible.

21 Which substance might potentially be the most effective experimental chemical to apply to the stump of a rat leg accidentally severed, in an attempt to cause regeneration of the lost tissue?

   a *Xenopus* egg extract
   b extract from the lost leg
   c rat brain extract
   d sea urchin blastula extract
   e rat embryo endoderm cell extract

For further thought

1 Recall from Chapter 16 that the oocyte of vertebrates undergoes a long period of development. Review the processes that take place during this period and explain their significance in the early developmental stages of the organism.

2 What would happen if more than one sperm fertilized an egg? How is this prevented?

3 Explain the various factors in embryonic development that cause cells to have different characteristics despite their identical genetic content.

4 Contrast the properties of mosaic and regulative eggs. What type of egg do human beings have?

5 In what ways does the metamorphosis of amphibians resemble that of an insect that undergoes complete metamorphosis? How is it different?

6 Aging has a profound effect on cerebral function in human beings. From midlife on there is a continuing loss of neurons from the brain and parts of the spinal cord. Suggest factors that may be involved in this aging process. (Remember neurons normally do not divide after very early childhood.)

7 Discuss the similarities and differences between the processes of embryonic development and regeneration.

Chapter 18

EVOLUTION

KEY CONCEPTS

1 Modern evolutionary theory is based on two concepts: that the genetically determined characteristics of living things change with time, and that this change is directed by natural selection.

2 Evolutionary change means change in allelic frequencies (or genotypic ratios) in populations of organisms, not in individuals.

3 By natural selection, we mean nonrandom reproduction, or, more specifically, reproduction that is to some degree correlated with genotype.

4 The evolutionary raw material for natural selection is genetic variation in the population.

5 Natural selection can act on genetic variation only when it is expressed as phenotypic variation.

6 Evolutionary change is not automatic; it occurs only when something disturbs the genetic equilibrium. Mutation pressure and selection pressure are always disturbing this equilibrium, and migration and genetic drift may also do so.

7 Many characteristics are both advantageous and disadvantageous; the evolutionary fate of such characteristics depends on the algebraic sum of the separate selection pressures.

8 Adaptations are genetically determined characteristics that enhance an organism's chances of perpetuating its genes in future generations.

9 An adaptation need not be one hundred percent effective to give the individuals that possess it a significantly greater chance of surviving to reproduce.

154 • CHAPTER 18

10 In the modern view, a species is a genetically distinctive group of natural populations that share a common gene pool and are reproductively isolated from all other such groups.

11 Divergent speciation usually begins when external barriers separate two population systems geographically; as the two populations evolve independently, they accumulate differences that will lead in time to the development of intrinsic isolating mechanisms.

12 Another kind of speciation—speciation by polyploidy—is especially important in plants but rare in animals.

13 Divergent evolution—the evolutionary splitting of lineages into many separate lineages—has occurred very frequently; it has produced the immense diversity of living things.

14 One task of systematic biology is to discover the relationships among species and to reconstruct their evolutionary history, which modern classification systems attempt to reflect.

OBJECTIVES

After studying this chapter and reflecting on it, you should be able to

1 State the two fundamental concepts of modern evolutionary theory.

2 Explain why genetic variation must be expressed phenotypically to lead to evolutionary change, yet exclusively phenotypic variations are not raw material for evolution.

3 Give reasons why Lamarck's hypothesis of inheritance of acquired characteristics is not accepted today.

4 Determine allelic frequencies and set up a Punnett square to determine genotypic frequencies in a population.

5 Give the Hardy-Weinberg Law, and discuss its meaning for evolutionary theory.

6 Explain what natural selection means in modern evolutionary theory, and explain how natural selection may increase the frequency of a given allele in the gene pool of a population.

7 Contrast directional, stabilizing, and disruptive selection.

8 Contrast the roles of selection and mutation in directing evolutionary change.

9 Explain how natural selection can be both a creative and a conservative agent in evolution.

10 Define and give an example of heterozygote superiority, and explain how it favors polymorphism.

11 Define and give one advantage of polymorphism.

12 Define adaptation and give at least five different examples of adaptations, explaining how each is advantageous to the organism possessing it.

13 Explain how the Industrial Revolution led to a rapid evolutionary change in populations of the English pepper moth (*Biston betularia*). Include information on H. B. D. Kettlewell's experiments.

14 Distinguish between Batesian and Müllerian mimicry.

15 Define and give an example of each of the following: symbiosis, commensalism, mutualism, parasitism.

16 Discuss four adaptations of internal parasites, and explain why most well-adapted parasites do not kill or seriously harm their host species.

17 Define the term fitness and explain why altruistic behavior might, under some circumstances, increase fitness and hence not be maladaptive.

18 Discuss the possibility of the evolution of nonadaptive characters.

19 Define population, deme, race, and species.

20 Discuss intraspecific geographic variation and distinguish among the following terms: clinal variation, stepped clinal variation, and discordant geographic variation.

21 Explain the geographic-isolation model of speciation. Be sure to include mutation, recombination, gene pool, extrinsic isolating mechanism, intrinsic isolating mechanism, and natural selection.

22 Describe eight intrinsic isolating mechanisms.

23 Describe speciation by polyploidy and explain the difference between autopolyploidy and allopolyploidy.

24 Explain the term adaptive radiation, and discuss the evidence for this phenomenon, using the Galápagos finches as an example.

25 Give the relationship between the stability of the environment and the rate of evolution.

26 Explain why the modern definition of a species is difficult to apply to asexual organisms, fossil organisms, populations at an intermediate stage of divergence, and allopatric populations.

27 Give the sources of information used to determine phylogenetic relationships among organisms.

28 Explain why convergence poses a problem for the systematist.

29 Distinguish between the following pairs of terms: homologous and analogous, specialized and generalized, primitive and advanced.

30 List the classification hierarchy used today and explain how a species is named.

SUMMARY

Fundamental to the modern theory of evolution are two concepts—that the characteristics of living things change with time and that this change is directed by natural selection. Evolution is the change in the genetic makeup of a population in successive generations. There are always variations among members of a population; if there is selection against certain variants and for other variants, the overall makeup of the population may change with time.

Natural selection can act on genetic variation only when it is expressed phenotypically. Nongenetic variations and variations produced by somatic mutations are not evolutionary raw material. Somatic mutations do not affect the genotype of the germ cells. Lamarck's hypothesis of acquired characteristics is not tenable.

Population genetics is based on the concept of the *gene pool*, the sum total of all the genes possessed by all the individuals in the population. The frequencies of the various alleles of a given gene are used to characterize the gene pool. Evolution is a change in the allelic frequencies within gene pools.

Evolutionary change is not automatic; it occurs only when something disturbs the genetic equilibrium. According to the Hardy-Weinberg Law, both allelic frequencies and genotypic ratios remain constant from generation to generation in sexually reproducing populations if the following conditions for stability are met: large population (so there will be no *genetic drift*), no mutation (or else mutation equilibrium), no migration (i.e. no *gene flow* into or out of the population), and totally random reproduction. In reality, many populations are large, and some populations exist without migration, but the conditions of no mutation and random reproduction are never met in any population. Mutations are always happening and mutational equilibrium is rare; hence *mutation pressure* can cause slow shifts in allelic frequencies. Reproduction is never totally random; no aspect of reproduction is completely devoid of correlation with genotype. Nonrandom reproduction, or natural selection, is the universal rule.

Thus the Hardy-Weinberg Law describes the conditions under which there would be no evolution—and since these conditions cannot be met in nature, it follows that evolution is always occurring. Evolutionary change is a fundamental characteristic of the life of all populations.

The role of natural selection All populations are subject to *selection pressure*, which disturbs the genetic equilibrium. Even very slight selection pressures can lead to major changes in allelic frequencies over time. Changing environmental conditions can exert *directional selection*, which causes the population to evolve along a particular functional line.

Natural selection can be creative. Even in the absence of new mutation, selection can produce new phenotypes by combining old genes in new ways. Mutation, by contrast, is not usually a major directing force in evolution; its principal evolutionary role is to provide new variations upon which future selection can act.

Sometimes a population is subject to two or more opposing directional selection pressures. Such *disruptive selection* may divide the population into distinct groups.

Natural selection also plays a conservative role; it acts to preserve favorable gene combinations by eliminating less adaptive new combinations created by recombination and mutation. This sort of selection has been termed *stabilizing selection*.

Many characteristics have both advantageous and disadvantageous effects, and a single allele may influence multiple characteristics (*pleiotropy*). The evolutionary fate of such characteristics or alleles depends on whether the sum of all the various positive selection pressures acting on them is greater or less than the sum of the negative selection pressures.

Sometimes the effects of a given allele are more advantageous in the heterozygous than in the homozygous condition. In such cases, *heterozygote superiority* may lead to balanced polymorphic variation because it favors the retention of both alleles in the population. *Polymorphism*, the occurrence in a population of two or more distinct forms of a genetically determined character, can be advantageous because the different forms may exploit more completely the subdivisions of a variable environment.

Adaptations Adaptations are genetically controlled characteristics that enhance an organism's chances of perpetuating its genes in succeeding generations. Adaptations can be structural, physiological, or behavioral; genetically simple or complex; highly specific or general.

The flowering plants depend on external agents (e.g. wind, birds, insects) to carry pollen from the male to the female parts of the plant. The flowers are adapted in shape, structure, color, and odor to their particular pollinating agent. The plants and their pollinators have evolved together, each becoming more finely tuned to the other's peculiarities. Such evolutionary interaction is called *coevolution*.

Defensive secretions, *ambiguous body orientation*, and *cryptic* (concealing) appearances are all adaptations that help animals escape predation. Kettlewell's experiments on the light and dark forms of pepper moths (*Biston betularia*) showed clearly that those moths that most closely resemble their background have the best chance of escaping predation.

Some animals that are disagreeable to predators (because of sting, bad taste, or smell) have evolved warning (*aposematic*) coloration. They benefit by being gaudily colored and conspicuous because predators can easily learn to recognize and avoid them after one or two unpleasant encounters. The avoidance of aposematic insects may not depend solely on learning; this avoidance response may evolve in predators.

Another protective adaptation is *mimicry*: members of different species resemble (mimic) one another. There are two types of mimicry. In *Batesian mimicry*, an unprotected species resembles a distasteful species; the mimicry is based on deception. *Müllerian mimicry* involves the evolution of similar appearances by two or more distasteful species. This type of mimicry is advantageous to both because predators learn more easily to avoid both prey species.

Symbiosis Many organisms have evolved adaptations for *symbiosis*—for living together. There are three types of symbiosis: commensalism, mutalism, and parasitism. However, there are no sharp boundaries between them; they grade into one another.

Commensalism is a relationship in which one species benefits while the other neither benefits nor is harmed. The advantages the commensal species receives from its host frequently include shelter, support, transport, and/or food. Often it is difficult to determine whether a relationship is commensalism or *mutualism*, in which both species benefit.

Parasitism is a symbiotic relationship in which one species benefits while the other species is harmed. External parasites live on the outer surface of their host, while internal parasites live inside the host's body. Internal parasitism is usually marked by more extreme specializations than external parasitism; these include structural degeneracy (evolutionary loss of structures), resistant body walls, a complex life cycle, and large reproductive potential. Internal parasites usually tend to evolve towards greater specificity, both towards their host and towards the part of the host's body they inhabit. Over time the host and the parasite undergo coevolution, eventually reaching a dynamic balance in which both can survive without serious damage. Most long-established host-parasite relationships are balanced ones.

Altruistic behavior Many social animals exhibit *altruistic behavior*, behavior that may reduce the personal reproductive success of the individual exhibiting the behavior while increasing the reproductive success of others. *Kin selection* is one explanation for the evolution of altruistic behavior. Behavior that benefits the altruist's kin may enhance the altruist's *inclusive fitness* (the sum of the individual's own reproductive success plus the success of its relatives devalued in proportion to their genetic distance). In some instances reciprocity may be an explanation; an altruistic act by one individual may increase the chances that the other individual will repay in kind. In other cases, an act that seems altruistic actually serves selfish ends.

Nonadaptive characters Some biologists have suggested that many characteristics found in high frequencies in populations lack any survival value; they are selectively neutral or *nonadaptive*. However, it should be remembered that an adaptation need not be one hundred percent effective; an adaptive value so slight as to be undetectable to us may be great enough to result in evolutionary change. Just because we cannot see the adaptive significance of a character does not mean it has none.

A nonadaptive character could possibly increase in frequency in a small population through genetic drift. It is also possible that a nonadaptive character may be an incidental effect of a pleiotropic gene.

The intense directional selection for certain traits in human-induced evolution has often led to characteristics that are nonadaptive or even maladaptive; human intervention has altered the balance between the selection pressures that would operate in nature.

Species and speciation A *population* of sexually reproducing organisms is a group of individuals that share a common gene pool. A *deme* is a small local population; demes are usually temporary units of population that intergrade with other such units.

The existence of discrete clusters of living things that can be called species has long been recognized, but the concept of what a species is has changed many times. In the modern view, a *species* is a genetically distinctive group of natural populations (demes) that share a common gene pool and that are *reproductively isolated* from all other such groups.

The vast majority of plant and animal species show geographic variation, most of which probably reflects differences in selection pressures resulting from local environmental conditions. Gradual variation correlated with geography is called a *cline*. An abrupt shift in a character constitutes a stepped cline; the populations on the two sides of the step may be designated as *races* or *subspecies*. These are groups of natural populations within a species that differ genetically and that are partly isolated from each other reproductively because they have different ranges.

Divergent speciation is the process by which one ancestral species gives rise to two or more descendent species, which grow increasingly unlike as they evolve. The initiating factor is usually geographic separation; if a population is divided by some physical or ecological barrier the separated (*allopatric*) populations will no longer be able to exchange genes. In time, the populations will evolve in different directions because they start out with different gene frequencies, they experience different mutations, and they are exposed to different environmental selection pressures. Eventually, the populations may become genetically so different that they develop *intrinsic isolating mechanisms*—biological characteristics that prevent effective interbreeding should they again become *sympatric*.

Many intrinsic isolating mechanisms act by preventing effective mating. The two populations may be so specialized ecologically that they cannot become sympatric, or they may occupy different habitats; they may breed at different times, be behaviorally isolated, or be physically unable to mate. Even if mating occurs, fertilization may not take place or the embryo may not survive. If hybrids are born, they may be inviable or sterile.

The members of two closely related populations may be able to breed and produce fertile offspring. If the hybrids are as well adapted as the parents, the populations will fuse back together. If the hybrids are less well adapted than the parents, there will be strong selection pressure favoring forms of intrinsic isolation that prevent wrong matings, and the two populations will diverge more rapidly (*character displacement*) until mating between them is impossible.

Among plants there is another process of speciation—speciation by *polyploidy*. Polyploidy is the occurrence in cell nuclei of more than two complete sets of chromosomes. In autopolyploidy there is a sudden multiplication of the number of chromosomes, usually as a result of nondisjunction in meiosis. In allopolyploidy there is a multiplication of the number of chromosomes in a hybrid between two species.

Divergent evolution—the evolutionary splitting of species into many separate descendent species—results in an *adaptive radiation*. The rate of evolutionary divergence is not always constant; when conditions change rapidly and organisms have new evolutionary opportunities available to them, they may undergo a rapid evolutionary burst. Adaptive radiation like that of Darwin's finches on the Galápagos Islands helps account for the tremendous diversity among living things on the earth today.

The modern definition of a species is hard to apply to asexual organisms, fossil organisms, populations in intermediate stages of evolution, and allopatric populations.

The concept of phylogeny One of the tasks of systematic biology is to discover the relationships among species and to trace the ancestors from which they are descended. Reconstructing the evolutionary history, *phylogeny*, of any group of organisms entails considerable speculation.

The morphology of the adult and embryo, combined when possible with information from the fossil record and from life histories, has traditionally been the basic source of data for phylogenetic hypotheses. Comparative physiology, comparative behavior, and comparative ecology have also supplied valuable information. More recently, techniques have been developed to compare the proteins and DNA of different species. The systematist's task is to gather together all that is known about the organisms he studies, and to try to reconstruct a picture of the organisms and their interrelationships.

One problem in interpreting the similarities among organisms is *convergence*. Organisms that are not closely related may come to resemble each other because they occupy similar habitats and adopt similar environmental roles. Systematists must try to determine whether the similarities are *homologous* (inherited from a common ancestor) or merely *analogous* (similar in function and often in superficial structure) before speculating about the degree of relationship between the organisms.

It is also necessary in reconstructing phylogeny to consider development in time. Is a characteristic *primitive* (older and more like the ancestral condition) or *advanced* (less like the ancestral condition)? These are often confused with the terms *specialized* (adapted to a narrow way of life) and *generalized* (broadly adapted). Generalized characters are likely to be more primitive, and specialized ones more advanced.

Classification The classification system used today attempts to encode information about the organism's evolutionary history. It uses a hierarchy of categories in which each category (taxon) is a collective unit containing one or more groups from the next-lower level. The principal categories are: kingdom, phylum, class, order, family, genus, and species. In the modern system of nomenclature each species is given a name consisting of two Latin words; the first names the genus to which the species belongs, the second the individual species designation.

QUESTIONS

Testing recall

Determine whether each of the following statements is true or false. If false, correct the statement.

1 For evolution to occur in a population, reproduction must be totally random.

2 Genetic drift has more evolutionary impact on small populations than on large ones.

3 Somatic mutation does not provide raw material for evolution.

4 Many genes are simultaneously subject to both positive and negative selection pressures.

5 Polymorphism is the occurrence in a population of two or more distinct forms of a genetically determined character.

6 Cryptically colored insects are usually distasteful to predators.

7 Industrial pollution caused the English pepper moth (*Biston betularia*) to absorb coal dust and become dark.

8 Two unrelated sympatric beetle species look very much alike. Both produce a noxious chemical that makes them taste bad to potential avian predators. This is probably an example of Müllerian mimicry.

9 In a large population with random mating, mutational equilibrium, and no migration, a disadvantageous allele will gradually disappear.

10 Selection of heterozygotes can result in an increase in the frequency of lethal recessive alleles.

11 Pleiotropy provides one possible explanation for evolution of nonadaptive characteristics.

12 A selection pressure of 0.001 is high enough to produce significant evolutionary change.

13 By definition, no hybridization between two closely related species is possible.

14 When two or more closely related species are sympatric, differences in their courtship displays often play an important role in preventing hybridization between them.

15 Character displacement is more likely to be observed in two related bird species from different islands than in two such species from the same island.

16 Two good species can never produce viable hybrids.

17 Two races (or subspecies) of the same species cannot be sympatric for long without fusing.

18 Speciation by polyploidy is much more common in animals than in plants.

19 Strictly speaking, the modern definition of species cannot be applied to asexual organisms.

20 The wings of birds and the wings of butterflies are examples of convergently evolved structures.

21 The modern system of scientific naming of plants and animals dates from publication of Darwin's *Origin of Species*.

The following important terms are sometimes confused by students. Distinguish between the terms within each of the following groups.

22 adaptive – nonadaptive – preadaptive

23 genetic drift – founder effect

24 directional selection – disruptive selection – stabilizing selection

25 Batesian mimicry – Müllerian mimicry

26 cryptic appearance – aposematic appearance

27 symbiosis – commensalism – mutualism – parasitism

28 parasite – predator

29 individual fitness – inclusive fitness

30 species – subspecies

31 intrinsic isolating mechanism – extrinsic isolating mechanism

32 sympatry – allopatry

33 autopolyploidy – allopolyploidy

34 homologous – analogous

35 specialized – generalized

36 primitive – advanced

37 convergence – divergence

Testing knowledge and understanding

Choose the one best answer.

For questions 1–3, use the information below.

Two pirates and three Polynesian beauties settled on a previously uninhabited tropical island. All five of these settlers had brown eyes, but one man carried the recessive allele for blue eyes (his genotype was *Bb*).

1 When the island population reaches 20,000 individuals, about how many people would you expect to have blue eyes if you assume a Hardy-Weinberg equilibrium for the eye-color alleles (which would admittedly be very improbable)?

a 0
b 20
c 200
d 1000
e 2000

2 No new vagabonds arrive on the island and no one leaves this tropical paradise. If the number of blue-eyed individuals differs greatly from the answer to the previous question, this could be due to a combination of which factors?

a gene flow and genetic drift
b gene flow and natural selection
c genetic drift and natural selection
d gene flow and nonrandom mating

3 Twenty percent of this island population of 20,000 has type AB blood, but the population our original five settlers left had only 1 percent type AB individuals. This is most likely an example of

a founder effect.
b natural selection.
c selective mutation.
d gene flow.
e nonrandom mating.

4. Ten percent of the genes for coat color in a rabbit population are albino (b) while 90 percent code for a black coat (B). What percentage of the rabbits are heterozygous if the Hardy-Weinberg assumptions hold true?

   a 90 percent
   b 81 percent
   c 18 percent
   d 9 percent
   e 1 percent

5. A large population of a certain species of freshwater fish lives in South America. No close relatives of this species are known. Suppose you could somehow cause all mutations to cease in this population and prevent all immigration into the population and emigration from it. Which one of the following statements *best* expresses the probable future of the population?

   a All evolution will promptly cease because without mutation there will be no raw material for evolution.
   b The population will begin to deteriorate after four or five generations because of excessive inbreeding that will result from the absence of immigration and emigration.
   c The population will continue to evolve for a long time as selection acts on the variability produced by recombination.
   d Major evolutionary changes will continue in the population because of genetic drift.
   e Although the population will cease to evolve, it may survive for a long time if the environment remains constant.

6. The theory of natural selection postulates that

   a in each generation, all the individuals well adapted for their environment live longer and produce more progeny than the less well adapted individuals.
   b the deaths of individuals occur completely at random with respect to the physical environment.
   c the deaths of individuals occur completely at random with respect to their genotypes.
   d to at least a small extent the survival and reproductive success of individuals depend upon the extent to which they are genetically adapted to their environment.
   e most deaths of individual organisms occur soon after fertilization, as a result of hereditary deficiencies.

7. Natural selection can best be defined as

   a survival of the best adapted individuals.
   b differential reproduction.
   c differential population growth rates.
   d enhanced survival of those individuals best adapted to attract mates.
   e the elimination of the weak by the strong.

8. A city was intensively sprayed with DDT in 1953 in an effort to control houseflies. The number of flies was immediately greatly reduced. Each year thereafter the city was again sprayed with DDT, but the flies gradually increased in numbers until ten years later they were almost as abundant as they were when the control program began. Which one of the following is the most likely explanation?

   a Flies from other areas moved in and replaced the ones killed by the DDT.
   b The few flies that were affected by DDT but survived developed antibodies to DDT, which they passed on to their descendants.
   c The DDT caused new mutations to occur in the surviving flies, and this resulted in resistance to DDT.
   d The DDT killed susceptible flies but the few that were naturally resistant lived and reproduced, and their offspring repopulated the area.

9. Evolution of Batesian mimicry depends upon

   a similar genetic potential inherited from the common ancestor of the model and the mimic.
   b brain-controlled deposition of pigment by the mimic in response to seeing the model.
   c evolution by the model and the mimic of similar adaptations to the same selection pressures.
   d greater predation on the non-mimicking individuals of the mimic species than on the mimicking individuals.
   e genetic drift as a result of very small population size.

10. Which would be the most likely agent of pollination for a flower with no scent and very small colorless petals?

    a honeybees
    b hummingbirds
    c beetles
    d wind
    e butterflies

11 What form of symbiosis best describes the relationship between a farmer and his flock of egg-producing hens?

  a predation
  b parasitism
  c commensalism
  d mutualism
  e competition

12 Biologists have had difficulty understanding the evolution of apparently altruistic behavior in animals, especially in cases where the altruist is unrelated to the individual toward whom the altruism is directed. Which one of the following statements best expresses the dominant current thinking on this question?

  a There is positive selection for altruism because the sacrifice of the altruistic individual may benefit the species as a whole.
  b Altruism is a natural consequence of the evolution of complex social systems in higher animals.
  c There is selection for altruism because it helps ensure that at least a few individuals of the group will survive to reproduce, even if the altruists themselves do not.
  d Altruism can be understood if one recalls that it is the population that evolves, not the individual.
  e Since selection acts on individuals, altruism between unrelated individuals could have evolved only if such behavior increased the fitness of the altruistic individuals themselves.

13 Gradual variation of a characteristic correlated with geography is an example of

  a a cline.
  b polymorphism.
  c phenotypic variation.
  d genetic drift.
  e gene flow.

14 Which one of the following isolating mechanisms is generally considered most important in *initiating* the process of speciation? (Exclude speciation by polyploidy.)

  a geographic isolation
  b behavioral isolation
  c developmental isolation
  d seasonal isolation
  e hybrid sterility

15 You carefully study populations of two very similar meadow mice, one from the Northeast and one from Texas. You want to know whether the populations belong to the same species or two different species. You could *most* confidently decide this if you could

  a show that the ranges of the two mice overlap without hybridization occurring.
  b bring the two types of mice into the laboratory and show that they can cross and produce viable offspring.
  c demonstrate that the natural ranges of the two types of mice are entirely allopatric.
  d show that there are statistically significant morphological differences between the two types of mice.
  e show that the northeastern mice live in wetter habitats than the mice of Texas.

16 Population A occurs only in Africa. The closely related population B occurs only in South America. Only a few minor morphological differences between A and B can be found. Population C, which is closely related to both A and B and is sympatric with B, differs more noticeably from B in several characters. In an attempt to determine whether A and B belong to the same or different species, individuals from the two populations were mated in the laboratory; they produced viable offspring. Populations A and B must be regarded as

  a members of the same species and subspecies.
  b members of the same species but of different subspecies.
  c members of different species.
  d uncertainly related because insufficient evidence is available to settle the matter.

In the evolutionary tree in the figure, each letter is a different species: C, D, E, and F live on different continents. Use the diagram to answer questions 17 through 19.

A ⟶ B ⟶ C, D, E, F

17 The evolution of species C, D, E, and F from B looks like an example of

  a adaptive radiation.
  b branching evolution.
  c convergent evolution.
  d polyploid speciation.

18 Species D and F are

  a sympatric.
  b allopatric.

19 Which organism (A, B, C, D, E, or F) is the most primitive?

20 Under natural conditions (i.e. without human intervention), the production of mules would probably tend to

   a weaken the intrinsic reproductive isolating mechanisms between horses and donkeys.
   b reinforce the isolating mechanisms between horses and donkeys.
   c result eventually in the formation of a single species from the two parental species (horses and donkeys).
   d decrease character displacement between horses and donkeys.

21 A biologist studies a large field in which hundreds of plants, all belonging to a single species, are growing. Plants of this species occur nowhere else in the state, and no other species of the same genus are known to science. The biologist demonstrates that all the plants in the field are probably interfertile. Five years later the biologist again visits the field and notices a few plants that look slightly different from the others. Upon conducting cross-pollination experiments, he discovers that these unusual plants can cross with one another but not with the normal plants; they appear to be members of a second species closely related to the original one. The most likely explanation is that

   a a new mutant gene arose in some members of the original population, causing it to split into two species.
   b genetic drift in this relatively small population led to speciation.
   c habitat differences in two different parts of the field led to divergent adaptation that resulted in speciation.
   d the unusual plants are polyploids, which probably arose as a result of nondisjunction.
   e the original species first evolved polymorphism, and then each morph became a separate species.

22 One plant species has a diploid number of 8. A second plant species has a diploid number of 10. These two plant species hybridize and a single fertile hybrid is produced. This new species produces gametes which probably have how many chromosomes?

   a 4   d 18
   b 5   e 36
   c 9

23 Convergent evolution between two species would be most likely to occur as a result of

   a a series of identical mutations occurring in both species.
   b hybridization between the two species.
   c interbreeding by both species with members of a third species.
   d exposure of both species to similar selection pressures.
   e genetic drift between the species.

24 Which one of the following is a correct hierarchical sequence of taxonomic groups?

   a class – order – family – genus
   b order – class – family – genus – species
   c family – order – species – genus
   d class – phylum – order – family

25 The scientific name of the human species is properly written as

   a homo sapiens.   d *Homo Sapiens.*
   b *homo sapiens.*   e *Homo sapiens.*
   c Homo Sapiens.

26 Which of the following pairs represents two species?

   a German shepherd, poodle
   b honeybee worker, queen honeybee
   c Monarch butterfly, Viceroy butterfly
   d housefly larva, housefly adult
   e Eskimo, Peruvian Indian

For further thought

1 In a certain African tribe, 4 percent of the population is born with sickle cell anemia. What percentage of the tribe enjoys the selective advantage of the sickle cell gene by being more resistant to malaria than individuals who are homozygous for "normal" hemoglobin?

2 Approximately one child in 10,000 is born with PKU (phenylketonuria). What is the frequency of this allele in the population? What is the frequency of the normal allele? What percent of the population are carriers of this trait?

3. Artificial selection can alter the distribution of phenotypes in a population. Suppose the distribution of phenotypes within a population looked like this:

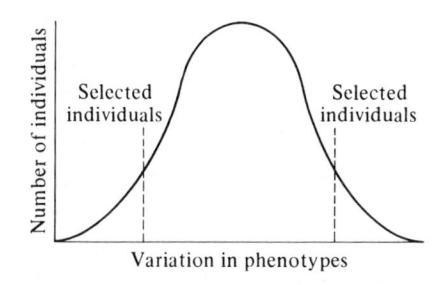

Disruptive selection was then practiced— only those organisms at both extremes of the distribution were chosen for breeding; those in the central portion were rejected. Draw a curve showing the expected phenotypic distribution of the F_1.

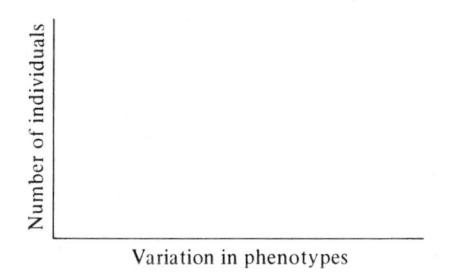

4. The introduction of DDT to control mosquitoes in the 1940s brought rapid control over the mosquitoes carrying the malarial pathogen. Today, resistance to DDT has developed in many mosquito populations and is reducing the effectiveness of malarial control programs. Explain how it was possible for the mosquitoes to evolve DDT resistance in such a short time.

5. Several small songbirds in different families have evolved similar "seeeee" alarm calls. What mechanisms might be involved in the evolution of the similar alarm calls?

6. The behavior of the Mexican Jay has been studied extensively. These birds are communal breeders. Nests are built by pairs; there are usually two active nests in the flock. Most of the members bring food to the nestlings. The parents bring about half of the food. The other half is brought by altruistic helpers: nonbreeding adults whose nests have failed, and immature birds. Suggest reasons for the evolution of this altruistic behavior.

7. Children born with Tay-Sachs disease die within two or three years of birth. The disease is inherited as an autosomal recessive. Considering the strong selection pressure against individuals with the disease, explain why the allele persists in the population.

8. The fossil record shows that the few surviving species of horseshoe crabs have existed unchanged for the past 200 million years. How do you account for the fact that there has been essentially no evolution of the species over this time span?

Chapter 19

ECOLOGY

KEY CONCEPTS

1 Many natural populations show an initial period of exponential growth at low densities, followed by a deceleration in growth at higher densities, and an eventual leveling off as the density approaches the carrying capacity of the environment. Such populations are primarily limited by influences that provide feedback control because they depend at least partially on the density of the population itself.

2 The density of other populations grows exponentially, but then falls precipitously as a result of density-independent limiting factors before reaching carrying capacity.

3 Predation, parasitism, competition (both intraspecific and interspecific), emigration, and physiological mechanisms are density-dependent influences on population growth.

4 The niche is the functional role and position of an organism in the ecosystem. The more the niches of two different species overlap, the more intense the competition between them.

5 A species does not exist as an isolated entity; it is always interacting in a variety of ways with other species in the community to which it belongs.

6 The movement of energy and materials within the community knits the community together and binds it with the physical environment as a functioning system.

7 Radiant energy from the sun is the ultimate energy source for life on earth. This energy is captured by the producers and passed on to the consumers.

8 Energy is constantly drained from an ecosystem as it is passed along the links of the food chain. Energy flow is always noncyclic.

9 The water and mineral components of the biosphere cycle through the ecosystems; the same materials are used over and over again, and can be passed around the ecosystems indefinitely.

10 The trend of most ecological successions is toward a more complex and stable ecosystem in which less energy is wasted and hence a greater biomass can be supported without further increase in the supply of energy.

11 Most ecological successions eventually reach a climax stage that is more stable than the stages that preceded it. Its more complex organization, larger organic structure, and more balanced metabolism enable it to buffer its own physical environment to such an extent that it can perpetuate itself as long as the environment remains essentially the same.

12 A limited number of regional climax formations called biomes can be recognized; the type of biome is determined by the most common climax of the region.

13 The distribution of biomes is a consequence of climate, physiography, and other environmental factors within each geographic area.

14 Since its intrinsic rate of increase always exceeds the carrying capacity of the environment, a population is always under pressure to expand its niche or to extend its range. To spread successfully into a new area, a species must have the physiological potential to survive and reproduce there, an ecological opportunity, and physical access.

15 The distribution of life today can only be understood by combining knowledge of present conditions with evidence from the fossil record and with geological evidence of the past configurations of the earth's land masses and their past climates.

OBJECTIVES

After studying this chapter and reflecting on it, you should be able to

1 Define and give the relationships between populations, communities, ecosystems, and biosphere.

2 Give two methods biologists use to estimate population densities.

3 Distinguish between uniform, clumped, and random distributions, and indicate the conditions under which each occurs, and which one is the most common.

4 Draw a graph of an exponential growth curve and a logistic growth curve, and write the equation for each. Define all terms in the equations. Indicate which curve is more representative of real populations.

5 Define zero population growth and explain how this condition is reached.

6 Draw an exponential growth curve with sudden crash, and list factors that might cause a crash.

7 Draw a graph showing the type I, type II, and type III survivorship curves, and indicate which curve is most common in natural populations.

8 Differentiate between density-dependent and density-independent limitations on population growth.

9 Distinguish between K strategists and r strategists, and give three general characteristics of each.

10 Discuss and give an example of the ways in which parasitism, predation, intraspecific competition, emigration, and physiological and behavioral mechanisms can act as density-dependent limits on populations.

11 Explain how destroying the balance between predator and prey in a population can upset the ecology of an area.

12 Explain the term "ecological niche," and relate this concept to the competitive exclusion principle and to the principle of limiting similarity.

13 Give the three possible results of intense interspecific competition.

14 Define each term and use it correctly: gross primary productivity, net primary productivity, producers, decomposers, primary consumers, secondary consumers, trophic level, food chain, food web.

15 Explain why the distribution of productivity within an ecosystem can always be represented as a pyramid, and why there are seldom more than four or five levels in a food chain.

16 Describe the pyramid of biomass and the pyramid of numbers and explain why these pyramids do not apply to all populations.

17 Outline the water, carbon, and phosphorus cycles. Indicate the role of microorganisms.

18 Describe the nitrogen cycle, including a discussion of the processes of nitrogen fixation, nitrification, and denitrification.

19 Discuss the effect of increased levels of phosphate ions in freshwater ecosystems, and give three important sources of these ions.

20 Define and give an example of the process of biological magnification.

21 Describe how the structure of the soil and its properties influence the availability of water, oxygen, and minerals to plants.

22 Discuss, using an example, the relationship between species diversity and complexity and community stability. Describe the effect of human intervention in biological communities.

23 Describe the process of ecological succession and its causes. Distinguish between primary and secondary successions.

24 Give five trends that characterize many successions.

25 Define the term "climax community," and contrast the monoclimax hypothesis with the more modern view of the climax community.

26 List the major biomes of the world; for each, give the principal characteristics of the area, some representative animals and plants, and the climatic factors that influence it.

27 Distinguish between each of the following pairs of terms: benthic division and pelagic division; euphotic zone and aphotic zone; neritic province and oceanic province.

28 Discuss the three major conditions that influence a species' success at colonization.

29 Describe how the Law of the Minimum and the Law of Tolerance are related to the distribution of species.

30 State the relationship between the size of an island and its species diversity.

31 Explain why some species become extinct after arriving in a new area.

32 List five means by which organisms are dispersed from one area to another.

33 Outline the main features of the continental drift concept, and indicate how it helps explain the present geography of life.

34 Discuss the similarities and the history of the two island continents, and indicate how they are reflected in the diversity of their biota.

35 Name the four biogeographic regions of the world continent, and explain why the biota of these regions are more alike than those of the island continents. Indicate what barriers separate these regions from one another.

SUMMARY

Ecology is the study of the interactions between organisms and their environment. There are three higher levels of organization: *populations*, groups of individuals belonging to the same species; *communities*, units composed of all the populations living in a given area; and *ecosystems*, the sum total of the communities and their physical environments considered together. The various ecosystems are linked to one another by biological, chemical, and physical processes. The entire earth is a true ecosystem; this global ecosystem is called the *biosphere*.

Populations as units of structure and function The distribution of individuals within an area may be uniform, random, or clumped. *Uniform* and *random* distributions are relatively rare and occur only where environmental conditions are fairly uniform. A uniform distribution results from intense competition or antagonism between individuals; a random distribution occurs when there is no competition, antagonism, or tendency to aggregate. *Clumping* is the most common distribution because environmental conditions are seldom uniform, reproductive patterns favor clumping, and animal behavior patterns often lead to congregation. The optimum density for population growth and survival is often an intermediate one; undercrowding can be as harmful as overcrowding.

All organisms have the potential for explosive growth; under ideal conditions their growth curve would be *exponential*. The equation for an exponential growth curve is $I = rN$ where I is the rate of increase of the population, r is the *intrinsic rate of increase* of the population (average birth rate − average death rate) and N the number of individuals in the population at a given moment. If r is positive, the population will grow at an ever accelerating rate.

The exponential growth of many real populations begins to level off as the density approaches the *carrying capacity* (K) of the environment. Such a growth curve is called an S-shaped or *logistic growth curve* and results from a changing ratio between births and deaths. After the curve has leveled off, births and deaths are in balance and the population has *zero population growth*. This occurs because environmental limitations become increasingly effective in slowing population growth as the population density rises. When the density approaches the carrying capacity, the limitation becomes severe. A *density-dependent limitation*, $(K − N)/K$, is one whose density is determined by the density of the very population it helps limit. The equation for the logistic growth curve is

$$I = r\left(\frac{K-N}{K}\right)N$$

At low population densities, population growth is exponential; the rate of increase reaches its maximum at the inflection point of the curve. When the population exceeds the carrying capacity, $(K − N)/K$ becomes negative, so I becomes negative, and the population decreases.

The populations of many small short-lived animals, or those living in variable environments, go through a period of exponential growth, followed by a sudden crash (boom-and-bust curve). The crash occurs before the populations reach the carrying capacity; it is due to a *density-independent* limitation such as weather or other physical environmental factors. The operation of such a limitation does not depend on the density of the organisms.

In addition to the birth rate and death rate, the potential life-span, the average life expectancy, and the average age of reproduction are important determinants of the makeup of a population. Determining the mortality rates for the various age groups in the population gives a survivorship curve. A type I curve is one in which all the organisms live to old age and die quickly; a type II curve shows a constant mortality rate at all ages; and a type III curve is typical of populations where the mortality among the young is very high, but those who survive the early stages tend to live for a long time. In nature, high mortality among the young is the rule.

Population regulation The regulation of population density in organisms with boom-and-bust curves is primarily due to the work of density-independent limitations. The maximum density achieved before the decline is primarily a function of the organisms' high reproductive rates. Since these organisms have evolved high intrinsic rates of increase (i.e. high r), they are called r *strategists*.

Organisms with S-shaped growth curves, whose population limitation is primarily density-dependent, are called K *strategists*; the maximum density for population stability is determined largely by the environment's carrying capacity. The fitness of these organisms depends on their ability to exploit the limited environmental resources efficiently.

Both parasitism and predation usually influence the prey (or host) species in a density-dependent manner. In general, the density of predators or parasites fluctuates in direct proportion to the changes in the density of the prey. In stable predator-prey systems, predation may be beneficial; it helps to regulate the population size of the prey and prevent it from outrunning its resources. Predator-prey relationships, like long-established host-parasite relationships, tend to evolve toward a dynamic balance in which predation is a regulatory influence in the life of the prey species, but not a real threat to its survival.

Intraspecific competition is one of the chief density-dependent limiting factors. As population density rises, competition for limited environmental resources becomes increasingly intense and acts as a brake on population growth.

Interspecific competition can also act as a density-dependent limitation. The more the niches of the species overlap, the more intense the interspecific competition. *Niche* refers to the functional role and position of an organism in an ecosystem. Every aspect of an organism's existence helps define that organism's niche. Niche is an abstract concept and can never be fully measured. According to Gause's principle, the *competitive exclusion principle*, two species cannot for long simultaneously occupy the same niche in the same place.

The more similar two niches are, the more likely it is that both species will be competing for a limited resource. According to the principle of limiting similarity, there is a limit to the amount of niche overlap compatible with coexistence. Extinction of one of the competing species, range restriction, character displacement, or a combination of the last two, will usually be the outcome of intense interspecific competition.

In some animals, crowding induces physiological and behavioral changes that result in increased emigration from the crowded region. Dense populations often experience disease epidemics. Numerous laboratory experiments with mice indicate that crowding induces hormonal changes which reduce the reproductive rate; the importance of this mechanism in nature is unclear.

The economy of ecosystems The flow of energy and materials knits a given community together and binds it with the physical environment as a functioning system. Almost all forms of life obtain their high-energy organic nutrients, directly or indirectly, from photosynthesis. The total amount of energy bound into organic matter by photosynthesis is called *gross primary productivity*; *net primary productivity* is the amount left after subtracting the amount the plant uses in respiration. Heterotrophs obtain their energy by consuming green plants, other heterotrophs, or the dead bodies or wastes of other organisms.

The *food chain* is the sequence of organisms, including the *producers* (autotrophic organisms), *primary consumers* (herbivores), *secondary consumers* (herbivore-eating carnivores), and *decomposers*, through which energy and materials may move in a community. In most communities the food chains are completely intertwined to form a *food web*. The successive levels of nourishment in the food chains are called *trophic levels*. The producers constitute the first trophic level, the primary consumers the second, the herbivore-eating carnivores the third, and so on. Since many species eat a varied diet, trophic levels are not hard-and-fast categories.

At each successive trophic level there is loss of energy from the system; only about 10 percent of the energy at one trophic level is available for the next. The distribution of productivity within a community can be represented by a *pyramid of productivity*, with the producers at the base and the last consumer level at the apex. In general, the decrease of energy at each successive trophic level means that less biomass can be supported at each level; thus many communities show a *pyramid of biomass*. Some communities also show a *pyramid of numbers*: there are fewer individual herbivores than plants, and fewer carnivores than herbivores.

The endless cycling of water to earth as rain, back to the atmosphere through evaporation, and back to earth again as rain maintains the freshwater environments and supplies water for life on land. The *water cycle* is also a major factor in modifying temperatures and in transporting chemical nutrients through ecosystems.

Carbon cycles from the inorganic reservoir to living organisms and back again. Carbon dioxide in the atmosphere or in water is converted into organic compounds by photosynthesis; the resulting organic compounds may be released as CO_2 by respiration or be consumed by animals or decomposers. Eventually the carbon will be released as CO_2 and the cycle will begin again. Human activities have increased the CO_2 level in the atmosphere; eventually this may lead to a change in the earth's temperature since the heat radiated from earth is absorbed by atmospheric CO_2 and radiated back to warm the earth in a "greenhouse effect."

Biological *nitrogen fixation* by microorganisms, some of them living symbiotically in root nodules, provides most of the usable nitrogen for the earth's ecosystems. The microorganisms reduce atmospheric N_2 to NH_3, which is often in the form NH_4^+. *Nitrification* may then occur; different groups of bacteria convert NH_4^+ to NO_2^- and then to NO_3^-, which may be absorbed by roots and converted into organic nitrogen. When the plant dies, nitrogen is returned to the soil as NH_3, which can be recycled. Some bacteria carry out *denitrification*, converting NH_3, NO_2^- or NO_3^- into N_2 gas.

Phosphorus also cycles from the inorganic reservoir to living organisms and back again. Phosphate rock dissolves slowly and becomes available to plants, which pass it to animals. Some is excreted by animals; the rest is released from organic compounds when the organism dies. Huge quantities of phosphates flow into aquatic environments in runoff water. Sewage, detergents, and fertilizer runoff have greatly increased the phosphate level, which accelerates the *eutrophication* (aging) process of lakes.

Modern industry and agriculture have been releasing vast quantities of chemicals into the environment. Some, such as mercury, are harmless when released, but are made toxic by microorganisms. Some chemicals (e.g. DDT) show *biological magnification*. When ingested, these persistent chemicals are retained in the body and tend to become increasingly concentrated as they are passed up the food chain to the top predator.

The properties of soils—their particle sizes, amount of organic material, and pH, among others—determine how rapidly water and minerals move through the soils and how available to plants they will be. The proportions of clay, silt, and sand particles help determine many of these soil properties. Decaying organic material (*humus*) promotes proper drainage and aeration.

A complex equilibrium exists between the ions free in the soil water and those adsorbed on clay and organic particles. Many factors, acidity in particular, can shift this equilibrium; for example, acid rain, a by-product of air pollution, damages the soil. The plants and animals living in or on the soil also profoundly affect soil structure and chemistry. The cutting down of forests, poor farming practices, and irrigation have ruined many soils.

The biotic community The species comprising a community interact with each other and with the physical environment. The biotic community they form can be considered a unit of life, with its own characteristic structure and functional interrelationships. Species diversity and complexity of interactions influence community stability. A simple community responds violently to a disturbance but often recovers quickly. The complex community responds less dramatically but may continue to show effects over a longer period. A diverse physical environment seems to favor community stability. Monoculture, pollution, and other human activities have increased community instability by decreasing species diversity and structural complexity.

Ecological succession is an orderly process of community change involving the replacement, over time, of the dominant species within a given area by other species. Succession results in part from the modification of the habitat by the organisms themselves or by physiographic changes, and in part from differences in dispersal and growth patterns among species. The species that predominate in the early stages tend to be rapid growers that are easily dispersed.

Successions in different places and at different times are not identical. The sequence of changes in *primary successions* (those occurring in newly formed habitats) is longer and slower than in *secondary successions* (those occurring in areas where previous communities were destroyed). The trend of most successions is toward a more complex and longer-lasting ecosystem in which less energy is wasted and hence a greater biomass can be supported without further increase in the supply of energy.

If no disruptive factor interferes, most successions reach a stage that is more stable than those that preceded it—the *climax community*. It will persist as long as the climate, physiography, and other environmental factors remain the same. Earlier ecologists supported the monoclimax hypothesis, which held that all succession in a given climatic region will converge to the same climax type. Modern ecologists, however, argue that since each species is distributed according to its own particular biological potentialities, climax has meaning only in relation to the individual site and its environmental conditions.

Most biologists recognize a number of major climax formations called *biomes*. The *tundra* is the northernmost biome of North America, Europe, and Asia. The subsoil is permanently frozen. There are many organisms on the tundra but relatively few species. South of the tundra lies the zone dominated by the coniferous forests, the *taiga*. More different species live in the taiga than on the tundra.

The biomes south of the taiga show much variation in rainfall and thus more variation in climax communities. The *deciduous forests* predominate in temperate zones with abundant rainfall and long, warm summers. Tropical areas with abundant rainfall are usually covered by *tropical rain forests*, which include some of the most complex communities on earth; the diversity of species is enormous. Huge areas in both the temperate and tropical regions are covered by *grassland* biomes. These occur in areas of low or uneven rainfall. Places where the rainfall is very low form the *desert* biomes. Deserts are subject to the most extreme temperature fluctuations of any biome type.

A series of different biomes can also be found on the slopes of tall mountains. Climatic conditions change with altitude, and biotic communities change correspondingly.

Aquatic ecosystems also vary with varying physical conditions. Oceanic ecosystems may be classified as the *benthic division* (the ocean bottom with all bottom-dwelling organisms) or the *pelagic division* (the water above the bottom with all the swimming and floating organisms). Another system distinguishes between an upper well-lighted *euphotic zone* and a deeper lightless *aphotic zone*. Still another possible distinction is between the *neritic province* above the continental shelves and the *oceanic province* of the main ocean basin. The most complex oceanic communities occur in the *littoral zone*, the shallow waters along the beach. These subdivisions are not fully analogous to terrestrial biomes since energy and materials flow between the different subdivisions.

Biogeography Since its intrinsic rate of increase always exceeds the carrying capacity of the environment, a population is always under pressure to expand its niche or to extend its range. A species must meet three conditions to spread into a new area: it must possess the *physiological potential* to survive and reproduce in that area; it must have the *ecological opportunity* to become established, and it must have *physical access* to the new area.

Colonization is possible only if the colonizers are already at least minimally preadapted to survive under the new environmental conditions. Climate and weather are important limitations on the distribution of species. Often the extremes of the weather limit the distribution. The same kinds of organisms will not live at all points within a given region since the microenvironmental conditions may vary. Other environmental factors are also important. Liebig's *Law of the Minimum* states that the growth of a plant will be limited by whichever required factor is most deficient in the environment. Shelford's *Law of Tolerance* states that the distribution of a species will be limited by that environmental factor for which the organism has the narrowest range of adaptability.

The colonizing species must have the ecological opportunity to become established in the new area; it must encounter little competition at first and find an available niche. Studies of island biogeography show an area-species rule—the greater the area, the greater the species diversity. Generally, the number of species doubles for every tenfold increase in area.

An organism must also have physical access to a new area. Organisms may be dispersed from one place to another by active locomotion or by passive transport (e.g. by air, water currents, birds, or mammals). The southeast Pacific island of Krakatoa constitutes a natural experiment in colonization of new territory and illustrates different dispersal mechanisms. Studies of island biogeography indicate that the more distant an island is from a major source of new colonists, the lower the species diversity at its equilibrium state. The geological or ecological barriers that separate two regions will be more effective barriers to some species than to others.

Distribution of living organisms To understand the present geography of life, we must combine knowledge of present conditions with evidence from the fossil record and with geological evidence of past configurations of the earth's land masses and their climates. Geological evidence indicates that 225 million years ago all the earth's land masses were combined in a single supercontinent called *Pangaea*. Pangaea broke up into a northern supercontinent called *Laurasia* and a southern one called *Gondwana*. Soon Gondwana broke up; India drifted to the north and the African-South American mass separated from the Antarctic-Australian mass. Later, each of these masses split. The division of Laurasia into North America and Eurasia was one of the last to occur. As the continents moved, their climates changed, altering the distribution of organisms. In addition, fossil evidence indicates that the earth's climate has undergone many changes over time.

The biota (flora and fauna) of the *Australian region* is most unusual; many species, particularly the mammalian fauna, that are common in Australia exist nowhere else. Most of the ecological niches that are filled by placental mammals on other continents are filled by marsupials in Australia. The unusual biota can be explained by the long isolation of Australia from the other continents.

The South American continent, the *Neotropical region*, has also been an island continent through much of its history, but its nearness to North America and its recent connection via the Central American land bridge have given it a more diverse biota.

Europe, Asia, Africa, and North America form the World Continent; their biotas are relatively similar. The World Continent can be divided into the *Nearctic* (North America), *Palaearctic* (Europe, northern Asia), *Oriental* (southern Asia), and *Ethiopian* (Africa south of the Sahara). After the division of Laurasia, North America and Eurasia were connected through much of their history by the Siberian land bridge.

QUESTIONS

Testing recall

Match the following descriptive phrases with the terms they best define. Use each letter only once.

a abiotic
b biosphere
c climax community
d community
e competition
f decomposers
g ecological niche
h ecosystem
i eutrophication
j K
k mortality rate
l population
m producers
n pyramid of energy
o r
p succession
q trophic levels

1 autotrophs in a community
2 all organisms in a given place at a given time
3 functional role of an organism in its community
4 progressive change in the plant and animal life of an area
5 intrinsic growth rate of a population
6 a major determiner of population density
7 global ecosystem
8 successive levels of nourishment in a food chain
9 carrying capacity
10 group of individuals belonging to the same species
11 end of every food chain
12 stable stage of succession
13 a density-dependent limitation on population growth
14 aging process in a lake
15 sum total of the physical features and organisms in a given area

Determine whether each of the following statements is true or false. If it is false, correct it.

16 Clumped spacing of members of a population is more common than uniform or random spacing.
17 A population with more offspring per generation will always have a faster growth rate (r) than a similar population with fewer offspring per generation.
18 The house fly, which has a short life-span and produces a large number of eggs, could be considered a K strategist.
19 Predation usually acts as a density-independent limiting factor on populations.
20 The pyramid of productivity holds true for all populations.
21 The pyramid of numbers holds true for all populations.
22 In a normal biological community, there will be less usable energy at the carnivore trophic level than at the herbivore trophic level.
23 In most biological communities there are more top predators than primary consumers.
24 Most flowering plants obtain their nitrogen in the form of nitrate.
25 The term "niche" designates the part of an organism's habitat that includes physical environmental factors, such as temperature, humidity, and soil pH.
26 Carnivores are more likely to have high concentrations of DDT in their tissues than comparable herbivores in the same ecosystem.
27 Food webs usually become simpler as ecological succession proceeds.
28 The species composition of the community in a given area changes markedly during the course of ecological succession.
29 Energy utilization in a climax community is usually more efficient than in a pioneer community.
30 The subsoil of the taiga is permanently frozen.
31 A temperate deciduous forest usually has fewer plant species than a taiga forest.
32 No producers are found in the benthic division in the aphotic zone of the ocean.
33 The establishment of biotic communities on Krakatoa was an example of secondary succession.
34 Islands far from the mainland tend to have fewer species than islands of similar size located closer to the mainland.
35 Italy is in the Nearctic region.

Testing knowledge and understanding

Questions 1-7 refer to the following situation.

A New York State farmer stocked his farm pond with 1000 bluegill fish fingerlings. Bluegills usually reproduce first as yearlings and regularly thereafter. The farmer recorded the number of fish each year for the next ten years. He obtained the following data:

Year	Number of fish
Stocked	1000
1	750
2	580
3	600
4	750
5	1200
6	1400
7	1460
8	1440
9	1450
10	1460

1. Plot these data on the graph.
2. Why did the population decline during the first two years?
3. What kind of growth curve is seen in years two through ten?
4. Would you consider the bluegill a K-strategist or an r-strategist?
5. Mark the point on the curve where the rate of increase is greatest.
6. What factors might be involved in slowing the population growth from the sixth year on?
7. At what point should the farmer begin fishing if he wants to avoid overexploiting the fish population?

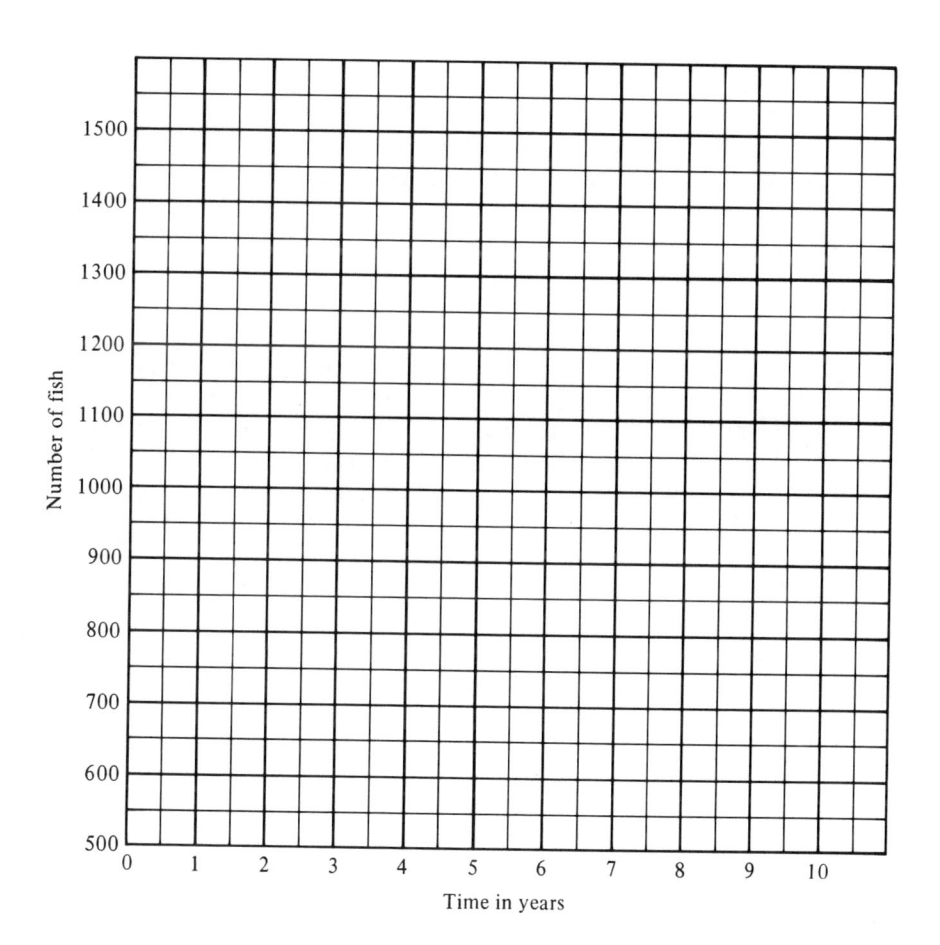

Choose the one best answer.

8 Which one of the populations below will show the greatest population increase in the next year?

   a population A with 200,000 individuals when $r = .020$
   b population B with 500,000 individuals when $r = .040$
   c population C with 2 million individuals when $r = .008$
   d population D with 10 million individuals when $r = .002$
   e population E with 30 million individuals when $r = .001$

9 A certain fish species lives in large lakes. It feeds on small insects and other invertebrates. Spawning occurs in late spring, when males and females congregate in shallow water and engage in brief courtship displays, after which they release gametes in great clouds. The adult fish then swim back to deeper water. Young fish become sexually mature when they are three years old. Which of the following survivorship curves is most likely to apply to this species?

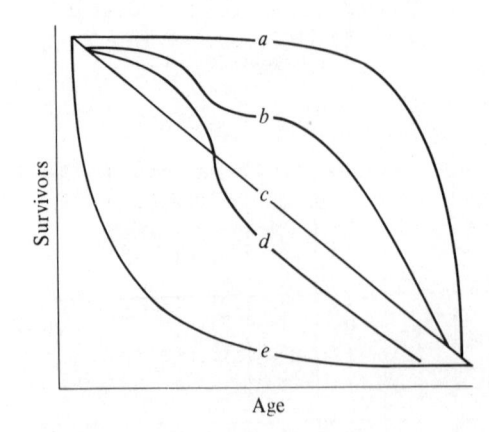

10 Which item below cannot help regulate animal populations?

   a seasonal changes in weather
   b polymorphism
   c competition
   d predation
   e social hierarchies

11 Which one of the following would be least likely to act as a density-dependent limiting factor on a population of insects?

   a parasitism
   b competition
   c predation
   d unfavorable climate
   e physiological changes induced by crowding

12 The diagram shows a particular food web. Each letter represents a different species. Arrows indicate the flow of energy and materials. Which of the following would have the greatest total biomass?

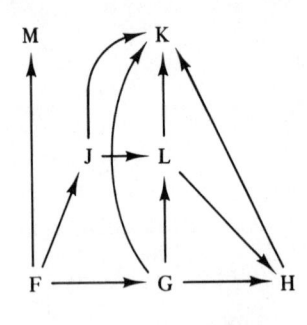

   a F
   b J + G
   c K
   d K + M
   e H

13 Which one of the species is probably carnivorous?

   a J   c L
   b M   d G

14 Which species is a decomposer?

   a F   d K
   b G   e L
   c H

The arrows in questions 15-20 indicate the direction of energy flow. For each question, evaluate the validity of the direction of the arrow. If the arrow points the correct way, the answer is a; if the arrow is incorrect, the answer is b.

15 autotroph ⟶ heterotroph

16 herbivore ⟶ carnivore

17 dead carnivore ⟶ saprophyte

18 respiration ⟶ photosynthesis

19 primary consumer ⟶ secondary consumer

20 producer ⟶ decomposer

21 The second law of thermodynamics helps explain why in any balanced biological community (over an extended time) there must be

   a a greater mass of photosynthetic plants than of heterotrophic organisms.
   b a greater mass of decomposers than of autotrophic organisms.
   c a greater mass of green plants than of minerals in the soil in which they grow.
   d more carnivores than herbivores to insure that the green plants will not all be destroyed.
   e a greater mass of secondary consumers than of primary consumers.

22 In a natural, stable biotic community, which ecological pyramid would be least likely to be inverted?

   a biomass
   b numbers
   c production
   d body size

23 Which stage in the nitrogen cycle provides the nitrogen source for use by carnivores?

   a ammonia (NH_3)
   b nitrate ions (NO_3^-)
   c nitrate ions (NO_2^-)
   d atmospheric nitrogen (N_2)
   e amino acids

24 Water, oxygen, and minerals are most available to plants growing in soils that are called

   a sands.
   b loams.
   c clays.
   d silts.

25 Which one of the following statements is *not* true?

   a Theoretically, the ecological niche of any given species is determined by an almost infinitely large number of different factors.
   b According to the so-called pyramid of biomass, there will usually be less biomass in the carnivores than in the herbivores in a food chain.
   c The reproductive potential of any species of plant or animal far exceeds the actual reproductive rate.
   d Severe interspecific competition often leads to extinction, range restriction, or evolutionary divergence.
   e Food chains almost always begin with some sort of unicellular autotrophic organism.

26 If two sympatric species occupy very similar niches under natural conditions, one would expect

   a the species to hybridize.
   b intense interspecific competition to occur.
   c extensive interspecific cooperation to occur.
   d the carrying capacity of the environment to be reduced.
   e a mutualistic symbiosis to develop.

27 Which one of the following is *not* a trend in ecological succession?

   a an increase in the number of trophic levels
   b an increase in productivity
   c an increase in community stability
   d a decrease in nonliving organic material
   e an increase in species diversity

28 Which one of the following is *least* likely to be true of an ecological succession?

   a The species composition of the community changes continuously during the succession.
   b The total number of species rises initially, then stabilizes.
   c The total biomass in the ecosystem declines after the initial stages.
   d The total amount of nonliving organic matter in the ecosystem increases.
   e Although the amount of new organic matter synthesized by the producers remains approximately the same after the initial stages, the percentage utilized at the various trophic levels rises.

29 The climax stage of a biotic succession

   a persists until the environment changes significantly.
   b changes rapidly from time to time, seldom remaining at any stage for more than a decade or so.
   c is the first stage in the reclamation of land from a lake bottom.
   d is a stage in which the dicot plants are always dominant.

30 Which ecosystem is the most unstable?

   a Sahara desert
   b African grassland
   c Alpine tundra
   d Montana wheat field
   e Costa Rican tropical rain forest

31 The climax formation (biome) of much of the state of Kansas is

 a tundra.
 b grassland.
 c tropical rain forest.
 d taiga.
 e deciduous forest.

32 The growing season is generally shortest in the

 a coniferous forest.
 b deciduous forest.
 c taiga.
 d tundra.
 e tropical rain forest.

33 Which one of the following statements is *not* true?

 a Deserts often display extreme daily variations in temperature.
 b Epiphytes are very common in tropical rain forests.
 c The subsoil of the tundra is permanently frozen.
 d The dominant trees of the taiga are birch and maple.
 e Vast numbers of waterfowl nest on the tundra.

34 Boston is in

 a the temperate deciduous biome of the Nearctic region.
 b the taiga biome of the Nearctic region.
 c the temperate deciduous biome of the Palaearctic region.
 d the taiga biome of the Palaearctic region.
 e the temperate deciduous biome of the Neotropical region.

35 Which one of the following is *not* a requirement for a colonizing species?

 a an available niche in the new area
 b a favorable climate
 c physical access to the new area
 d physiological potential to survive in the new area
 e an active means of locomotion for dispersal

36 Which biogeographic regions are on island continents?

 a Oriental and Ethiopian
 b Nearctic and Palaearctic
 c Nearctic and Neotropical
 d Australian and Neotropical
 e Australian and Oriental

37 The present distribution of plant and animal life can best be explained by assuming that the earth's land masses have drifted from one place to another. According to this concept of continental drift

 a South America and Australia were connected through much of geologic time.
 b North America and Europe were connected throughout much of their geologic history.
 c India split away from Asia and drifted southward.
 d the entire Oriental region originally belonged to the same supercontinent as South America and Africa.
 e the separation of Europe from Asia occurred early in geologic history.

For further thought

1 The population of Brazil in 1978 was estimated to be 113,100,000. If the birth rate was 37.1/1000 and the death rate was 8.8/1000, what was *r* for this population? What will the approximate population be at the end of 1979? Contrast the annual population growth of Brazil with that of the People's Republic of China, where the population in 1978 was estimated to be 950 million, the birth rate was 26.5/1000, and the death rate was 10.3/1000. Which country will show the greatest population increase in the next five years if the growth rates remain unchanged?

2 In the food chain in Figure 19.5 in your textbook, the fox feeds on rabbits, squirrels, mice, and seed-eating birds, all of which feed on plant material. How many square meters of plant material are required to support one fox if the net primary productivity of the plant material is 8,000 Kcal/m^2/year? Assume the fox's daily caloric requirement is 800 Kcal, and that only 10 percent of the energy at one trophic level can be passed on to the next. If the fox were to feed only on insect-eating birds (see Figure 19.5), how many square meters of plant material would be required?

3. Trace the route that a molecule of CO_2 might follow as it cycles through the ecosystem. (Include at least four organisms in the pathway.)

4. Calcium ions are required nutrients for both plants and animals. The reservoir for calcium is in rocks. Design a calcium cycle, including at least three organisms in the cycle.

5. As the world's human population continues to grow, the problem of feeding the earth's population becomes increasingly difficult. What are some of the environmental consequences of attempting to feed so many people?

Chapter 20

THE ORIGIN AND EARLY EVOLUTION OF LIFE

KEY CONCEPTS

1 Life arose spontaneously from nonliving matter under the conditions prevailing on the early earth; from these beginnings all present life on earth has descended.

2 All the events now hypothesized in the origin of life and all the known characteristics of life seem to fall well within the general laws of the universe; no supernatural event was necessary to the origin of life on earth.

3 There was no abrupt transition from "nonliving" prebionts to "living" cells; the attributes associated with life were acquired gradually.

4 The first cells were probably procaryotic; eucaryotic cells may have originated from a symbiotic union of several ancient procaryotic cell types.

5 Living organisms, once they arose, changed their environment and so destroyed the conditions that made possible the origin of life.

6 The classification of living organisms into kingdoms is to some degree arbitrary; each classification system has its advantages and disadvantages.

OBJECTIVES

After studying this chapter and reflecting on it, you should be able to

1 Explain how modern ideas of spontaneous generation differ from those of early biologists.

2 Describe the formation of the earth and give two models for the formation of the early atmosphere.

3 Describe the formation of the small organic molecules in the early ocean.

4 State Oparin's hypothesis and explain how Miller's experiment supports it.

5 Explain how polymers may have formed in the ancient seas.

6 Describe coacervate droplets and proteinoid microspheres and indicate ways in which they resemble living cells.

7 Give two alternative models for the origin of the first cells.

8 Discuss the role that competition is thought to have played in the evolution of the various biochemical pathways.

9 Discuss possible reasons why the evolution of photosynthesis destroyed the conditions that made possible the origin of life.

10 Evaluate the possibility of life on other planets.

11 Explain why fossils from the Precambrian are so scarce and why there was such an enormous increase in life forms in the Cambrian.

12 Discuss the evidence for the endosymbiotic model for the origin of the eucaryotic cell.

13 Give the rationale for the classification of living things into five kingdoms.

SUMMARY

Scientists today believe that life could and did arise spontaneously from nonliving matter under the conditions that prevailed on the early earth, and that all present earthly life has descended from these beginnings. The basis for the current theory of the origin of life was stated by A. I. Oparin in 1936.

The solar system was probably formed between 4.5 and 5 billion years ago from a cloud of cosmic dust and gas. As the earth condensed, a stratification took place; heavier materials moved toward the center and lighter substances were concentrated at the surface. Eventually the lighter gases of the earth's first atmosphere escaped into space. The intense heat in the interior of the earth drove out various gases through volcanic action; these formed a second atmosphere.

The atmosphere of the early earth was quite different from today's oxidizing atmosphere. Two principal models of the early atmosphere have been proposed. The first states that the early atmosphere contained much H_2 and was a reducing atmosphere. Thus nitrogen was present as ammonia, oxygen as water vapor, and carbon as methane. The second model assumes that the atmosphere was made up primarily of gases that occur in present-day volcano outgassings: H_2O, CO, CO_2, H_2. In either model, the early atmosphere contained almost no O_2.

As the earth's crust cooled, the water vapor condensed into rain and began to form the oceans, in which gases from the atmosphere and salts and minerals from the land dissolved. Ultraviolet radiation, lightning, and heat could have provided the energy for chance bonding of these substances to form the organic building-block molecules. Experimental evidence shows that such abiotic synthesis of organic compounds can occur. These organic compounds could have accumulated slowly in the seas over millions of years since they would not have been destroyed by oxidation or decay.

Some investigators feel that the organic material in the oceans became sufficiently abundant for chance polymerizations to occur; others suggest that various concentrating mechanisms were necessary for polymerization of the building-block molecules.

Oparin speculated that *coacervate droplets* then formed; these are clusters of macromolecules surrounded by an orderly shell of water molecules. Such droplets have a definite internal structure and can absorb substances selectively. Fox proposes, instead, the formation of *proteinoid microspheres*, which exhibit many properties of living cells. Vast numbers of such prebiological systems may have arisen in the seas. Some may have contained favorable combinations of materials and grown in size; new droplets could have formed upon fragmentation. Somehow, the nucleotide sequences in nucleic acids came to code for the sequence of amino acids in protein. Those droplets with particularly favorable characteristics developed into the first cells.

Alternatively, some biologists suggest that the first "living" things were self-replicating nucleic acids that slowly surrounded themselves with cytoplasm and a membrane.

The earliest organisms were heterotrophs that obtained energy from the nutrients available in the early ocean. As the nutrients disappeared, competition between the organisms must have increased. Natural selection favored any new mutation that enhanced an organism's ability to obtain or process food. Over time, various biochemical pathways evolved that enabled organisms to utilize different nutrients. The first form of metabolism using ATP was probably fermentation.

As the free nutrients were used up, some organisms evolved the ability to use another energy source—the sun. Cyclic photophosphorylation probably evolved first, then noncyclic photophosphorylation. From this time onward, life on earth depended on the activity of photosynthetic autotrophs. The oxygen released by algal photosynthesis converted the atmosphere to an oxidizing one. With free O_2 available, organisms could evolve efficient aerobic respiration. The oxygen also gave rise to a layer of ozone in the upper atmosphere that shielded the earth's surface from intense ultraviolet radiation.

Sincd life could have evolved spontaneously from nonliving matter on earth, it is also possible that it could have arisen elsewhere in the universe.

Evolution of the eucaryotic cell The oldest cellular fossils are of bacteria and are about 3.1 billion years old. Most authorities think the blue-green algae evolved from bacteria about 2.3 billion years ago. Few fossils of higher forms of life are found from the Precambrian, but they are abundant in the Cambrian. The Berkner-Marshall hypothesis suggests that this great increase in living forms resulted from the increase in atmospheric O_2 to about 1 percent of present levels. Later, when the O_2 reached about 10 percent of present levels, it cut the intensity of ultraviolet light reaching the earth sufficiently to permit life to move onto land.

The first fossils of eucaryotic cells are about 1.5 billion years old, but there is no direct evidence of their evolution. Several investigators think that the chloroplasts, mitochondria, and cilia may be modern descendants of ancient procaryotic cells that became obligate endosymbionts of other cells and have evolved in concert with their hosts ever since. A growing list of characteristics indicates that mitochondria and chloroplasts have many features in common with free-living procaryotic organisms. If this model is correct, the chloroplasts are probably derived from blue-green algae, and the mitochondria from aerobic bacteria. There is little evidence on the origin of the nuclear membrane and the endoplasmic reticulum.

The kingdoms of life The evolutionary relationships between the major groups of organisms are poorly known. Despite this ignorance, science continues to attempt to assign all living things to a few large categories called kingdoms or divisions. Whatever criteria are chosen, it is impossible to make a clean separation between the groups; these are artificial human categories imposed on nature.

The classification system used in your textbook recognizes five kingdoms: the Monera, Protista, Plantae, Fungi, and Animalia. The kingdom Monera includes the procaryotic organisms. It is believed that they gave rise to many eucaryotic lineages. Those that terminate at a unicellular, colonial, or very primitive multicellular level constitute the kingdom Protista. Three major lineages reached the higher multicellular level. Each exploits a different mode of nutrition—photosynthetic autotrophism is used by plants (kingdom Plantae), absorptive heterotrophism by fungi (kingdom Fungi), and ingestive heterotrophism by animals (kingdom Animalia).

QUESTIONS

Testing recall

Arrange the following steps in the evolution of life in chronological order.

1 competition

2 formation of the earth

3 prebionts

4 reducing atmosphere

5 oxidizing atmosphere

6 primitive procaryotic cell

7 abiotic synthesis of building-block molecules

8 biochemical pathways

9 O_2 revolution

10 great rains

11 polymerization

12 eucaryotic cell

13 photosynthesis

Testing knowledge and understanding

Choose the one best answer.

1. Which one of the following has been proven experimentally?

   a Life on earth originated by spontaneous generation.
   b Life on earth originated by special creation.
   c Complex organic molecules can be produced abiotically from methane, ammonia, and water vapor.
   d Spontaneous generation cannot occur today.
   e Eucaryotes evolved from procaryotes.

2. Which one of the following statements about current scientific thinking on the origin of life is false?

   a The first autotrophs were probably eucaryotic cells.
   b Given great expanses of time, raw materials, and energy, the basic building-block organic compounds could be synthesized in the absence of living cells.
   c The first living organisms must have depended on glycolysis and fermentation for their livelihood.
   d As living organisms evolved, they changed the environment so much that the conditions that made possible the origin of life no longer exist.
   e Coacervate droplets have some attributes of living cells; they are orderly arrangements of molecules with a membrane that can selectively absorb substances from the surrounding medium.

3. Which one of the following statements is false?

   a Coacervate droplets have a strong tendency toward formation of definite internal structure.
   b The chemical reactions in a coacervate droplet depend in part on the physicochemical organization of the droplet itself.
   c There is a definite interface between a coacervate droplet and the liquid in which it floats.
   d Coacervate droplets have a marked tendency to absorb and incorporate various substances from the surrounding medium.
   e The regular spatial arrangement of the molecules within a coacervate droplet probably reduces the catalytic activity of proteins in the droplet.

4. The first *cellular* organisms on earth were probably most like today's

   a eucaryotic cells.
   b viruses.
   c bacteria.
   d green algae.
   e Protozoa.

5. Competition probably became important at an early stage in the origin of life because

   a nucleic acids cannot duplicate themselves.
   b available organic materials were being consumed faster than new materials were being synthesized.
   c cells arose early in the sequence as a result of the colloidal characteristic of proteins.
   d the early proteins probably acted as catalysts and thus accelerated the various chemical reactions that were taking place.
   e the ability to carry on photosynthesis had become widespread.

6. Which of the following events led most directly to the decrease in the amount of ultraviolet radiation reaching the surface of the earth?

   a evolution of cyclic photophosphorylation
   b evolution of noncyclic photophosphorylation
   c development of a reducing atmosphere
   d evolution of heterotrophy
   e evolution of fermentation

7. Early in the evolution of life on earth, metabolic breakdown of energy-rich compounds must have been only by fermentation, because aerobic respiration

   a occurs only in animals, but the earliest organisms were plants.
   b could not occur before the evolution of blue-green algal or green-plant photosynthesis.
   c could not occur after the so-called oxygen revolution.
   d could not occur before the first tracheophytes developed.
   e can only oxidize carbohydrates, but the early nutrients were mostly of other types.

8 It has been proposed that mitochondria, chloroplasts, and flagella are modern descendants of primitive forms of procaryotic cells that took up residence within primitive cells and evolved independently there. Which one of the following is *not* evidence of this view?

   a Like procaryotic cells, mitochondria and chloroplasts have DNA that is not closely associated with protein.
   b The ribosomes in procaryotic cells, mitochondria, and chloroplasts are very similar chemically, and differ from the cytoplasmic ribosomes of eucaryotes.
   c Certain antibiotics inhibit the ribosomes in the mitochondria and chloroplasts but do not affect cytoplasmic ribosomes.
   d Mitochondria and chloroplasts are known to have their own genes and ribosomes and to conduct protein synthesis.
   e Mitochondria and chloroplasts can be cultured outside cells.

9 According to the five-kingdom system used in your textbook, the Protista includes

   a most unicellular eucaryotes.
   b the procaryotic organisms.
   c all unicellular organisms.
   d unicellular eucaryotes and the algae.
   e unicellular eucaryotes and the fungi.

10 The kingdom Monera includes: (Use the choices in question 9.)

11 Which one of the following is mismatched?

   a Fungi – absorptive heterotrophs
   b Animalia – ingestive autotrophs
   c Plantae – autotrophs
   d Monera – bacteria and blue-green algae
   e Protista – primarily unicellular or colonial eucaryotic groups

For further thought

1 Another hypothesis for the origin of life on earth suggests that life did not originate here; it was brought here by some extraterrestrial object such as a meteor. The earth is constantly bombarded by meteorites. Analyses indicate that some meteorites contain molecules (some hydrocarbons) characteristic of living systems. Comment upon this hypothesis.

2 Suppose you were given the following organisms to classify:

   a A unicellular organism that has chlorophyll and carries on photosynthesis. It lacks a cell wall and is highly motile.
   b A large multinucleated amoeboid organism that carries on phagocytosis. Its reproduction is plantlike; it forms groups of thin-walled spore cases similar to those of fungi.

In which kingdom would you classify each if you were using the five-kingdom system? How would you classify them in the four-kingdom system? Give the advantages and disadvantages of each system.

Chapter 21

VIRUSES AND MONERA

KEY CONCEPTS

1. Viruses are on the borderline between living and nonliving; they lack the metabolic machinery to make ATP and proteins, and they cannot reproduce themselves in the absence of a host, yet they have nucleic acid genes that encode information for their reproduction.

2. The cells of bacteria, blue-green algae, and the newly discovered Prochlorophyta are procaryotic; they lack a nuclear membrane and most other membranous organelles and thus are fundamentally different from all other living organisms.

3. The bacteria are an extraordinarily successful group of organisms; their extreme metabolic versatility and enormous reproductive potential have enabled them to survive in a wide variety of habitats.

4. Many bacteria are beneficial; all other organisms depend directly or indirectly on the activities of the bacteria.

5. The blue-green algae produce oxygen as a by-product of their photosynthesis; they probably initiated the oxygen revolution some 2.3 billion years ago.

OBJECTIVES

After studying this chapter and reflecting on it, you should be able to

1. Explain why it was not until 1935 that scientists were able to distinguish between viruses and bacteria.

2. Describe the structure of a virion.

3. Contrast the reproductive cycle of a bacteriophage with that of a virus that infects the cells of higher organisms.

4 Distinguish among the terms virion, vegetative virus, and provirus.

5 Distinguish between RNA transcriptase and reverse transcriptase.

6 Give three hypotheses for the origin of viruses.

7 Explain the role of interferon.

8 Describe a viroid.

9 State the difference between procaryotic and eucaryotic cells.

10 List the divisions of the Monera.

11 List four ways in which bacteria differ from viruses.

12 Draw a diagram of a typical bacterial cell, and give the three basic shapes.

13 Explain how the cell wall of bacteria differs from those of plants and fungi, and explain the significance of the cell wall in the life of bacteria.

14 Give the function of bacterial endospores.

15 Contrast the structure and mode of action of a bacterial flagellum with a eucaryotic flagellum.

16 Describe the process of reproduction in bacteria and explain how it differs from that of eucaryotic organisms.

17 Explain how genetic recombination occurs in bacteria.

18 Describe the three nutritive modes found in bacteria.

19 Differentiate among aerobes, obligate anaerobes, and facultative anaerobes.

20 List Koch's postulates.

21 Explain how pathogenic bacteria may harm the body of their host, and tell how the body resists the attacks of such pathogens.

22 Distinguish between a vaccine and an antiserum; active and passive immunity.

23 List the ways in which the Cyanophyta resemble bacteria, and the ways in which they differ.

24 Contrast the photosynthetic activity of the blue-green algae with that of bacteria and that of higher plants.

25 Explain why the blue-green algae can live in very inhospitable environments.

26 Give the possible evolutionary significance of the Prochlorophyta.

SUMMARY

Viruses In the late nineteenth century the first evidence of infectious agents so small that they could pass through fine porcelain filters was found. In 1935 the tobacco mosaic virus was crystallized, and the filterable viruses were recognized as different from bacteria.

Viruses are not cells; though viruses possess a few enzymes, they lack the metabolic machinery for energy generation, and they never have the ribosomes required for protein synthesis. The free virus particle, or *virion*, consists of a nucleic acid core covered with a protein coat (*capsid*). The nucleic acid can be DNA or RNA, double-stranded or single-stranded.

Viruses cannot reproduce themselves; the host cell manufactures new viruses using the genetic instructions provided by the virus. When bacteriophages reproduce, only the nucleic acid enters the host cell; in most other cases the entire virion enters. Once inside the cell, the nucleic acid provides the genetic information for the synthesis of new viral nucleic acid and proteins.

If the nucleic acid is DNA, it acts as a template for the synthesis of both mRNA and new viral DNA. If the nucleic acid is RNA, replication usually takes place in the cytoplasm; an enzyme, *RNA transcriptase*, uses the RNA as a template to synthesize new RNA. In some RNA viruses, the retroviruses, the RNA is transcribed instead into DNA in a reaction catalyzed by *reverse transcriptase*. The newly formed DNA then becomes integrated into the host's chromosomes.

Once new viral components have been synthesized by the host cell, they are assembled into new virions. Some virions are released by lysis of the host cell, others by an extrusion process in which the virus becomes enveloped in a piece of cell membrane.

Viruses can exist in three states: as free infectious virions, as viral nucleic acid directing replication in a host cell, and as provirus.

Three hypotheses of viral origin have been proposed. The first suggests that viruses are organisms that have reached the extreme of evolutionary specialization for parasitism; they have lost all cellular components except the nucleus. The second hypothesis suggests that modern viruses represent a primitive "nearly living" prebiont stage in the origin of life. The third hypothesis suggests that viruses are fragments of genetic material derived from other organisms.

Viruses cause many diseases in various organisms, and have been associated with some kinds of cancer. Most viral infections do not respond to antibiotic treatment. A protein called *interferon* promotes recovery from viral infections. Produced by the host's cells in response to an invading virus, it is released from infected cells and interacts with uninfected cells, giving them a resistance to infection.

A new kind of infectious agent called a *viroid* has recently been discovered. Viroids consist of a small amount of RNA with no protein coat. The viroid RNA may have some regulatory action in the host cell.

Monera The kingdom Monera includes three divisions: the Schizomycetes (bacteria), Cyanophyta (blue-green algae), and Prochlorophyta. The cells of these organisms are procaryotic.

Bacteria The division Schizomycetes includes a wide variety of forms that are now closely related. The bacteria are cellular; they have the metabolic machinery to generate ATP and to reproduce themselves; and they have ribosomes for protein synthesis. Bacteria occur in three basic shapes: spherical (*cocci*), rod-shaped (*bacillus*), or corkscrew (*spirilla*). Some bacteria remain together after cell division and may form clusters, chains, or other characteristic groupings.

The cell walls of bacteria and other monerans differ from those of plants and fungi; they are made of murein, a polymer of polysaccharide chains cross-linked by amino acids. Penicillin is toxic to growing bacteria because it inhibits murein formation. Since most bacteria live in a hypotonic environment, they would burst without a cell wall. A few very small parasitic bacteria, the mycoplasmas, do lack a cell wall. Differences in the cell wall chemistry of different bacteria are associated with differences in their staining properties; these can be used for identification. An envelope or capsule may surround the cell wall.

Some bacteria can form special resting cells called *endospores*, which enable them to withstand adverse conditions. When conditions improve, the endospores produce new bacterial cells.

Many bacteria are motile. Some have flagella, which are entirely different from the flagella of eucaryotic cells. Other bacteria move with a peculiar gliding motion.

Bacteria have enormous reproductive potential. The single circular chromosome composed of DNA is found in the *nucleoid* area of the cell; it replicates before division. Bacteria reproduce by *binary fission*, a type of nonmitotic cell division that produces two daughter cells exactly like the parent. Although this process is asexual, many bacteria do have some genetic recombination; new DNA can enter the cell by conjugation, transduction, or transformation.

Most bacteria are heterotrophic (either saprophytic or parasitic), but some are *chemosynthetic autotrophs* and others are *photosynthetic autotrophs*. The chemosynthetic bacteria oxidize certain inorganic compounds and trap the released energy. The photosynthetic bacteria, unlike higher plants, lack chlorophyll *a* (they have bacteriochlorophyll instead), do not use water as an electron source, and do not produce O_2. They possess only Photosystem I. The oxygen requirement of bacteria varies; some are *aerobic*, others are *facultative anaerobes*, and still others are *obligate anaerobes*.

While studying pathogenic bacteria, Robert Koch formulated the rules of procedure for proving that a particular microorganism is the cause of a particular disease. *Koch's postulates* are still used today.

Microorganisms cause disease symptoms by interfering with normal function, by destroying cells and tissues, or by producing poisons called *toxins*.

The human body resists the attacks of pathogenic organisms with the phagocytic action of white blood cells and with the production of antibodies. Some diseases can be prevented by immunization; the patient is injected with a vaccine or an antiserum. *Vaccines* consist of antigenic substances from the pathogen; these produce a long-lasting active immunity by stimulating the production of antibodies in the patient. An *antiserum* contains presynthesized antibodies against a specific antigen, which produce an immediate, short-term, passive immunity.

Beneficial bacteria far outnumber harmful ones. They play a vital role in the cycling of materials in the ecosystem, are important in many food and industrial processes, and are used to produce antibiotics.

188 • CHAPTER 21

Cyanophyta The Cyanophyta, or blue-green algae, are procaryotic unicellular or filamentous photosynthetic organisms. They exist singly, as small colonies, or as multicellular filaments. The cell walls are composed of murein. A layer of gelatinous material, the sheath, often surrounds the wall. These organisms reproduce by binary fission.

All blue-green algae possess chlorophyll *a*, the pigment found in higher plants, and generate O_2 as a by-product of photosynthesis. The pigments are located in membranous vesicles (thylakoids) but are not contained within chloroplasts.

Many blue-green algae can fix atmospheric nitrogen. Certain cells, the *heterocysts*, are specialized for nitrogen fixation. High levels of O_2 inhibit the processes of nitrogen fixation and photosynthesis.

The blue-green algae can live in freshwater, saltwater, and most terrestrial habitats. Some live mutualistically with fungi in lichens. Because they are photosynthetic and can fix nitrogen, their nutritional requirements are few and they can live in environments where no other organisms can survive.

Prochlorophyta The Prochlorophyta are procaryotic organisms with a pigment system like that of green algae and higher plants. They would seem more likely than the Cyanophyta to be the progenitor of chloroplasts in eucaryotic cells since they resemble eucaryotic chloroplasts much more than the Cyanophyta do.

QUESTIONS

Testing recall

Fill in the blanks.

A virus consists of an inner core of a single molecule of nucleic acid which may be (1) _____ or _____. Surrounding the nucleic acid is a coat of protein called the (2) _____. The viruses that attack bacterial cells are called (3) _____. They attach by their tail to the bacterial cell wall and the (4) _____ is injected into the host cell. The energy for this injection comes from the hydrolysis of (5) _____. The phage nucleic acid provides the genetic information for the synthesis of new viral (6) _____ and _____. The new virions escape by (7) _____ of the cell.

In plant and animal viruses, the (8) _____ enters the host cell. If the virus is an RNA virus, the RNA molecule can serve as a template for the synthesis of new RNA, provided the enzyme (9) _____ is present to catalyze the reaction. In the retroviruses, however, the RNA serves as a template for the synthesis of (10) _____. The enzyme (11) _____ is required for this reaction. This nucleic acid becomes integrated into the host's chromosome and is called a (12) _____. Once the new virions have been assembled they may be released by (13) _____ of the cell or by (14) _____ through the cell membrane.

Recently a new kind of infectious agent called a (15) _____ has been found. It consists only of RNA with no protein coat.

The kingdom Monera consists of the (16) _____, _____, and _____. The cells of all these groups lack a nuclear membrane and most other membranous organelles; they are called (17) _____ organisms.

The bacterial cell is surrounded by a cell wall made of (18) _____. Directly under the wall is the selectively permeable (19) _____. The genetic material is found in an area of the cell called the (20) _____. Some bacteria have (21) _____, which move the cell by their rotary motion. Under certain conditions some bacteria form (22) _____, which are resistant to destruction. Bacteria have (23) _____ within the cytoplasm to carry out protein synthesis. The three basic shapes of bacteria are (24) _____, _____, and _____. Bacteria reproduce by (25) _____.

Most bacteria show the (26) _____ mode of nutrition. Bacteria vary in their O₂ requirements; those that cannot survive in the presence of oxygen are called (27) _____. Most of these organisms obtain their energy by the process of (28) _____. Some bacteria contain chlorophyll but do not have (29) _____, which is the chief light-trapping pigment in higher plants.

Many bacteria cause disease. The rules of procedure for proving that a particular organism causes a disease are called (30) _____. Some bacteria harm their host by producing poisons called (31) _____. Many diseases can be prevented by inoculating a patient with a (32) _____ containing an antigen that stimulates the patient to produce (33) _____ against the disease-causing organism, conferring an (34) _____ immunity.

The Cyanophyta resemble the higher plants in possessing the light-trapping pigment (35) _____. This pigment is located in membranous sacs called (36) _____. The blue-green algae differ from the true plants in lacking (37) _____. Special cells called (38) _____ fix atmospheric nitrogen. The Cyanophyta divide by (39) _____. Recently new procaryotic organisms have been found that have a pigment system like that of higher plants. These organisms, called the (40) _____, may be the progenitors of the chloroplasts in eucaryotic cells.

Testing knowledge and understanding

Choose the one best answer.

1 Viruses are usually not classified as living organisms because they do not

 a contain DNA or RNA.
 b self-replicate.
 c possess chlorophyll.
 d possess nuclear membranes.
 e possess enzymes.

2 In which *one* of the following ways do viruses resemble cellular organisms?

 a They divide by mitosis.
 b They can undergo mutation.
 c They exhibit aerobic respiration.
 d They have an extensive endoplasmic reticulum.
 e They have a membrane composed largely of phospholipids.

3 Which one of the following statements is false?

 a Viruses have no metabolic machinery of their own.
 b Viral genes are composed only of DNA
 c Viruses cannot be cultured on artificial media.
 d Viruses will form crystals under certain conditions.

4 Which one of the following statements concerning viruses is false?

 a The nucleic acid in viruses may be DNA or RNA, either single-stranded or double-stranded.
 b Some viruses reproduce in the cytoplasm, others in the nucleus.
 c In both plant and animal cells, the entire virus particle usually enters the cell.
 d The nucleic acid in the virus codes for the proteins that make up the coat of the virus and for some enzymes.
 e Viruses are sensitive to most antibiotics, particularly penicillin and the sulfa drugs.

5 Reverse transcriptase synthesizes DNA on an RNA template. This enzyme is characteristic of

 a animal cells.   d certain DNA viruses.
 b bacterial viruses. e certain RNA viruses.
 c fungal cells.

6 Animal cells produce an antiviral protein known as

 a interferon. d penicillin.
 b viralase.   e capsid.
 c lysozyme.

7 Which entity consists only of DNA?

 a a bacteriophage
 b a provirus
 c a human chromosome
 d a bacterial spore
 e a viral capsid

8 The cells of Monera differ from those of eucaryotic organisms in many ways. Which of the following is *not* one of those ways?

   a Cells of Monera lack a nuclear membrane.
   b When chlorophyll is present in Monera cells, it is located in lamellae, but these are not contained in chloroplasts.
   c Cells of Monera lack mitochondria and an endoplasmic reticulum.
   d Cells of Monera lack ribosomes.
   e When the cells of Monera possess flagella, those flagella lack a 9 + 2 microtubular structure.

9 A bacterial cell does *not* possess

   a a cell wall.   *d* mitochondria.
   b DNA.   *e* ribosomes.
   c a plasma membrane.

10 Heterotrophic bacteria digest their food

   a extracellularly.   *c* in food vacuoles.
   b intracellularly.   *d* in lysosomes.

11 Which one of the following is never exhibited by any kind of bacteria?

   a conjugation   *d* spore formation
   b photosynthesis   *e* chemosynthesis
   c meiosis

12 In which case is *passive* immunity acquired?

   a A child contracts measles.
   b An adult gets a common cold.
   c Antibodies are passed from mother to baby in colostrum.
   d An adult gets a case of gonorrhea.
   e A child receives a tetanus "booster" shot.

13 Penicillin would be *least* effective against which disease listed below?

   a bubonic plague   *d* tuberculosis
   b gonorrhea   *e* strep throat
   c swine influenza

14 Blue-green algae are

   a heterotrophic procaryotes.
   b heterotrophic eucaryotes.
   c autotrophic procaryotes.
   d autotrophic eucaryotes.
   e protists.

15 Cyanophyta differ from Schizomycetes in

   a having no nuclear membrane.
   b possessing chlorophyll *a*.
   c having amino acids in their cell walls.
   d lacking lysosomes.
   e exhibiting no mitosis.

16 Blue-green algae do *not* have

   a chlorophyll *a*.
   b chloroplasts.
   c photosynthetic membranes.
   d autotrophic nutrition.
   e enzymes.

For further thought

1 Would you classify viruses as living or nonliving? Justify your answer.

2 Antibiotics are effective against many bacterial diseases, but most viral infections do not respond to treatment with them. However, a few antibiotics are somewhat effective against certain viruses. Below are listed four antibiotics and their mode of action. For each, determine whether it would be effective against bacteria and/or viruses and explain your answer.

   a penicillin – affects procaryotic cell wall synthesis
   b streptomycin – inhibits protein synthesis and causes misreadings of the mRNA code at the procaryotic ribosome
   c tetracycline – inhibits protein synthesis on the procaryotic ribosome
   d actinomycin – inhibits mRNA synthesis.

3 When penicillin was first introduced to treat syphilis and gonorrhea, it was extraordinarily effective, but certain physicians predicted that eventually it would become useless. In time, the spirochete causing syphilis did become resistant to penicillin, and recently a strain of *Gonococcus* (the organism causing gonorrhea) began to show resistance. Why have these organisms become resistant?

4 In the late 1940s it was discovered that feeding antibiotics to farm animals not only controlled infection but also stimulated growth. Today, half the antibiotics produced in the United States are used for animal feeds. Some scientists view this practice with alarm; in their opinion the widespread use of antibiotics is dangerous, and we may eventually return to the era of pre-antibiotics in the treatment of bacterial infections. Can you think of reasons for the scientists' concern?

5 Certain bacteria are essential components of the scheme of life. Discuss the role of bacteria in the cycling of inorganic nutrients. Could life survive if all bacteria were eliminated?

Chapter 22

THE PROTISTAN KINGDOM

KEY CONCEPTS

1 The Protista include the eucaryotic organisms that are primarily unicellular or colonial.

2 The Protista can be separated into three evolutionary lines: the animal-like, funguslike, and plantlike Protista. However, there are also many similarities among them.

3 The Mastigophora (Zooflagellata) appear to be the most primitive of all the Protozoa and may have given rise to the other protozoan groups, and possibly to the multicellular animals.

4 The plantlike protistans show a combination of plantlike and animal-like characteristics.

OBJECTIVES

After studying this chapter and reflecting on it, you should be able to

1 List the three major groups of the Protista, and relate them to the nutritional patterns of the Protista.

2 Explain why many biologists prefer to consider the protozoans as acellular organisms.

3 Give the general characteristics of the Protozoa, and list the five phyla.

4 List the four protozoan phyla discussed in your text, and give two distinguishing characteristics for each.

5 Describe the life cycle of the organism that causes African sleeping sickness and that of the organisms that cause malaria.

6 Describe the structure and reproductive pattern of the Protomycota.

7 Contrast the life cycle of the true slime mold with that of the cellular slime mold.

8 Give ways in which the Euglenophyta resemble plants and ways in which they resemble animals.

9 List the type of pigments, distinguishing characteristics, and economic importance of the Chrysophyta and Pyrrophyta.

10 Diagram the proposed evolutionary relationships among the members of the Protista.

SUMMARY

Recently the kingdom Protista has become restricted to those eucaryotic groups that are primarily unicellular or colonial. Although the protistans can be separated into three evolutionary lines, the animal-like, funguslike, and plantlike organisms, there are many similarities among them.

Animal-like Protista: the Protozoa Although protozoans are usually said to be unicellular, many biologists consider them *acellular* since they are far more complex than other individual cells. They live in a variety of aquatic and moist habitats, and exhibit great diversity of form. Most are heterotrophic. Reproduction is usually asexual but may be sexual. Many forms encyst under unfavorable conditions. The Protozoa are often divided into five phyla: Mastigophora, Sarcodina, Sporozoa, Cnidospora, and Ciliata.

The *Mastigophora*, or zooflagellates, move by flagella and appear to be the most primitive of the Protozoa. They may have given rise to other protozoan groups, and possibly to the multicellular animals. Most Mastigophora live as symbionts in the bodies of higher plants and animals. Some are parasites; for example, the genus *Trypanosoma* causes several diseases in human beings and domestic animals. One species causes African sleeping sickness, which makes large parts of Africa uninhabitable for humans.

The *Sarcodina* are the amoeboid Protozoa; they move by means of pseudopods. They are thought to be closely related to the zooflagellates because some zooflagellates undergo amoeboid phases, and a few sarcodines have flagellated stages. Several groups of sarcodines secrete hard calcareous or siliceous shells. The shells of two of these groups, the Foraminifera and Radiolaria, have formed much of the limestone and some of the siliceous rocks on the earth's surface.

The parasitic *Sporozoa* usually have a sporelike infective cyst stage in their complex life cycle. All are nonmotile. Malaria is caused by species of Plasmodium; the pathogen is transmitted from host to host by female *Anopheles* mosquitoes.

The *Ciliata* possess numerous cilia for movement and feeding. Their other organelles are greatly elaborated. The Ciliata differ from other protozoans in having a macronucleus and one or more micronuclei. The polyploid macronucleus controls normal cell metabolism; the micronuclei are concerned with reproduction and produce the macronucleus. Many ciliates reproduce occasionally by a sexual process called *conjugation*.

Funguslike Protista: Protomycota and Gymnomycota Two categories of fungal-type organisms can be recognized at the protistan level of complexity: the *Protomycota* (the true funguslike protists) and the *Gymnomycota* (the slime molds).

The Protomycota are saprophytes or parasites. Their haploid bodies may be a simple sac within the cell of a host, a sac with rootlike rhizoids, or a filamentous structure with a reproductive sac. During reproduction, flagellated cells are formed; they may function as gametes or as *zoospores*.

The slime molds are decidedly plantlike at some stages and animal-like at others. The life cycle of the *true slime molds* (division Myxomycota) proceeds from diploid multinucleate amoeboid plasmodium to stationary spore-producing plasmodium, to haploid spores, to flagellated gametes, to zygote, and back to the amoeboid plasmodium. The life cycle of the *cellular slime molds* (division Acrasiomycota) is quite different. The haploid spores form haploid free-living amoeboid cells. When food becomes scarce, some begin secreting cAMP, which causes the slime mold amoebae to aggregate and form a *pseudoplasmodium*, within which the individual cells remain distinct. Eventually a fruiting body develops and produces spores. The true slime molds and the cellular slime molds are probably not closely related.

The plantlike Protista Several groups of unicellular organisms show a combination of plantlike and animal-like characteristics. They are plantlike because many have chlorophyll and often a cell wall; they are animal-like in being highly motile.

The *Euglenophyta* are unicellular flagellates that lack cell walls. Like higher plants, their chlorophylls are *a* and *b*, but they store paramylum rather than starch. Many euglenoid species lack chlorophyll and are obligate heterotrophs; those that do possess chlorophyll are facultative heterotrophs. Euglenoids reproduce by longitudinal mitotic cell division; they have no sexual reproduction.

The yellow-green algae, the golden-brown algae, and the diatoms belong to the division *Chrysophyta*. These predominantly unicellular or colonial algae are shades of yellow or brown and possess chlorophylls *a* and *c* (no *b*). The walls of many are impregnated with silica or calcium. The undecomposed shells of diatoms form deposits of diatomaceous earth. The diatoms are an extremely important link in most freshwater and marine food chains; they are the most abundant component of the marine *plankton*. The phytoplankton are the principal photosynthetic producers in marine communities.

The second most important group of marine producers are the *Pyrrophyta*, or dinoflagellates. They are small, usually unicellular, organisms that typically possess two unequal flagella. The photosynthetic species possess chlorophylls *a* and *c*. Some species are luminescent, and some are poisonous. The so-called red tides that sometimes kill millions of fish are caused by species of red-pigmented dinoflagellates.

QUESTIONS

Testing recall

The chart lists the various groups of Protista and some important characteristics. After studying this chapter you should be able to fill in the chart using the directions below. The correctly completed chart will be a good study aid.

Column 1: Indicate whether the phylum or division contains organisms that are autotrophic, heterotrophic, or both.

Column 2: Indicate whether flagella are typically present or absent.

Column 3: List any mechanism of locomotion other than flagella.

Column 4: Indicate whether chlorophyll is present or absent. (If present give type(s).)

Column 5: List any other important or distinguishing characteristics.

Group	1 Autotrophic or heterotrophic	2 Flagella	3 Other means of movement	4 Chlorophyll pigments	5 Other distinguishing characteristics
Mastigophora					
Sarcodina					
Sporozoa					
Ciliata					
Protomycota					
Gymnomycota					
Euglenophyta					
Chrysophyta					
Pyrrophyta					

Testing knowledge and understanding

Choose the one best answer.

1. Which one of the following groups has no flagellated cells (except in the gametes)?

   a Chrysophyta
   b Mastigophora
   c Euglenophyta
   d Sporozoa
   e Pyrrophyta

2. Which one of the following is *not* true of the Protozoa?

   a Their food is usually digested in food vacuoles.
   b Movement is by pseudopodia, or by the beating of cilia or flagella.
   c Reproduction is usually by binary fission.
   d Many can encyst under unfavorable conditions.
   e Most freshwater forms have contractile vacuoles.

3. The organism that causes malaria belongs to the

   a Ciliata.
   b Mastigophora.
   c Sarcodina.
   d Cnidospora.
   e Sporozoa.

4. When a female *Anopheles* mosquito bites and transmits the *Plasmodium* into a human being

   a the sporozoites are discharged into the body.
   b the merozoites are released from the mosquito's salivary glands.
   c the gametes fuse and form an amoeboid zygote.
   d the injected sporozoites produce spores.
   e the amoeboid *Plasmodium* moves to the liver.

5. Which phylum of protozoans shows the most complex and well-developed organelles?

   a Mastigophora
   b Sarcodina
   c Ciliata
   d Sporozoa

6. Suppose you are given an unknown organism to identify. You find that it is unicellular and lacks a cell wall and locomotor organelles. It is an internal parasite. You conclude that it is most likely to be a member of the

   a Gymnomycota.
   b Schizomycetes.
   c Sporozoa.
   d Crysophyta.
   e Sarcodina.

7. The movement of slime molds most closely resembles that of

   a Ciliata.
   b Mastigophora.
   c Sarcodina.
   d Euglenophyta.
   e Sporozoa.

8. A true slime mold differs from a cellular slime mold in that the true slime mold

   a has both diploid and haploid stages in its life cycle.
   b has an amoeboid stage in its life cycle.
   c produces haploid spores.
   d secretes pulses of cAMP for aggregation.
   e reproduces only asexually.

9. All the animal-like and fungal Protista

   a are parasitic.
   b are heterotrophic.
   c have cells walls.
   d move by pseudopods.
   e reproduce by spores.

10. Which one of the following has both autotrophic and heterotrophic forms?

    a Euglenophyta
    b Sarcodina
    c Ciliata
    d Protomycota
    e Gymnomycota

11. The Crysophyta and Pyrrophyta resemble higher plants in possessing

    a chlorophyll *a*.
    b chlorophyll *b*.
    c starch as a storage product.
    d silica-impregnated cell walls.
    e two unequal flagella.

12. Which one of the following characteristics is unique to the Euglenophyta?

    a flagella
    b absence of a cell wall
    c paramylum reserve material
    d chlorophylls *a* and *b*
    e photosynthetic and nonphotosynthetic forms

13. Which one of the following is *not* true of the dinoflagellates?

    a Some can produce light.
    b Some cause "red tides."
    c They have chlorophyll and are photosynthetic.
    d They are an important component of the phytoplankton.
    e Their fossil remains form limestone and chalk.

For further thought

1 There is increasing evidence that persistent pesticides such as DDT may affect the activity of the marine phytoplankton. What ecological problems might result?

2 The World Health Organization lists malaria as the infectious disease that kills more individuals than any other. Malarial control programs center around the control of *Anopheles* mosquitoes. What stages of the life cycle of the *Plasmodium* take place in the mosquito? If all mosquito populations were exterminated, would malaria be eliminated?

3 In what ways do the plantlike Protista resemble higher plants? How are they different?

4 In Africa, humans and domestic cattle are parasitized by a species of *Trypanosoma* that causes the fatal African sleeping sickness in humans and nagana in cattle. Large areas of Africa are uninhabitable for humans and cattle because of these diseases. Although the native wild animals are infected with the trypanosomes, they show no symptoms of disease. The trypanosome is spread from host to host by the bite of the tsetse fly. At present, intensive research is going on with the intention of eventually eliminating the tsetse fly. What might be the ecological effect of this act?

Chapter 23

THE PLANT KINGDOM

KEY CONCEPTS

1 In the life cycle of the most primitive plants, the haploid stages are dominant; this apparently was the ancestral condition.

2 The evolution of most plant groups shows a tendency toward reduction of the gametophyte (multicellular haploid stage) and increasing importance of the sporophyte (multicellular diploid stage). The bryophytes represent the most conspicuous exception to this evolutionary trend; in them the haploid stages are dominant.

3 The evolution of the embryophytes is best understood in terms of adaptations for life in a terrestrial environment.

4 The angiosperms are the most successful land plants because they have evolved the most efficient adaptations for living and reproducing on land.

OBJECTIVES

After studying this chapter and reflecting on it, you should be able to

1 Distinguish between the Thallophyta and the Embryophyta.

2 State the probable evolutionary relationship between the green algae and the land plants.

3 Diagram the life cycle of a representative unicellular green alga such as *Chlamydomonas*. Indicate whether the dominant stage is haploid or diploid.

4 Trace the evolutionary changes in the volvocine series.

5 Distinguish among the terms isogamy, heterogamy, anisogamy, and oogamy.

6 Define the terms gametophyte and sporophyte, and explain alternation of generations. Diagram an algal life cycle showing alternation of generations.

7 Contrast the Phaeophyta and Rhodophyta with respect to: size, location, type of cell wall, storage product, and types of pigment.

8 Diagram and contrast the life cycles of *Ectocarpus* and *Fucus*.

9 Discuss the ways in which kelps such as *Laminaria* resemble land plants, and the ways in which they are different.

10 Discuss at least five problems plants face in a terrestrial environment.

11 Name at least three adaptations for terrestrial life that have evolved in the bryophytes, and give two reasons why these plants are restricted to a moist environment.

12 Diagram the life cycle of a moss or liverwort, and indicate the dominant stage.

13 Define and use correctly the following terms: homospory, heterospory, megaspore, microspore, antheridium, archegonium, sporangium, sporophyll.

14 List four distinctive characteristics of the Tracheophyta.

15 Trace the evolutionary advances of the tracheophytes from the psilopsids through the lycopsids, sphenopsids, pteropsids, and spermopsids. Indicate their order of appearance in the fossil record. Give at least one example of each group.

16 Diagram the life cycle of a typical fern and indicate which stage is dominant.

17 Explain why ferns and the more primitive tracheophytes are no better adapted for reproduction on land than the bryophytes.

18 Describe adaptations that probably contributed to the great evolutionary success of the Spermopsida.

19 Diagram the life cycle of a pine tree and discuss the ways in which this life cycle is more advanced than that of a typical fern.

20 Identify and give the functions of the parts of a typical flower.

21 Diagram the life cycle of an angiosperm plant.

22 Name at least four angiosperm adaptations for terrestrial life that are not found in the gymnosperms.

23 Explain how the angiosperms are adapted to solve each problem listed in objective 10.

24 Give at least five differences between monocotyledons and dicotyledons.

SUMMARY

The divisions of the plant kingdom have traditionally been separated into two groups: *Thallophyta* and *Embryophyta*. The thallophytes show little tissue differentiation, their reproductive structures are frequently unicellular and always lack a jacket of sterile cells, and embryonic development takes place outside the female reproductive structure. By contrast, the embryophytes show more tissue differentiation, their multicellular reproductive organs are surrounded by sterile jacket cells, and their early embryonic development occurs within the female reproductive organ.

In this classification system, the algal groups with many multicellular members are included in the Thallophyta; they belong in the divisions Chlorophyta, Phaeophyta, and Rhodophyta. The Embryophyta include two divisions, the Bryophyta and Tracheophyta.

Thallophyta The *Chlorophyta* (green algae) are probably the group from which the land plants arose. The green algae possess chlorophylls *a* and *b* and carotenoids. Many divergent evolutionary tendencies can be perceived within the Chlorophyta, including: the evolution of motile colonies, a change to nonmotile unicells and colonies, the evolution of coenocytic (multinucleated) organisms, and the evolution of multicellular filaments and three-dimensional leaflike thalluses (plant bodies with little tissue differentiation).

The unicellular green algae such as *Chlamydomonas* reproduce asexually by *zoospores* and sometimes show a simple form of sexual reproduction. There are no separate male and female individuals, and all the gametes are alike (*isogamy*). Isogamy is probably the primitive condition in plants. The zygote is the only diploid stage in the life cycle; dominance of the haploid stage is characteristic of most primitive plants.

The *volvocine series* is a series of genera showing gradual progression from the unicellular condition to an elaborate colonial organization. A number of evolutionary changes are seen in this series: a change from unicellular to colonial life; increased coordination of activity and increased interdependence among the cells; increased division of labor; and a gradual change from isogamy to *anisogamy* (gametes of two different sizes), to *oogamy* (motile male gamete, nonmotile female gamete).

The life cycles of many green algae include a multicellular stage. Often this stage is a filamentous thallus, which may be branching or nonbranching, depending on the species. Sexual reproduction is usually isogamous. The multicellular stage is haploid; the zygote is the only diploid stage.

Ulva is a multicellular green alga with an expanded leaflike thallus two cells thick. Its life cycle includes both multicellular haploid (*gametophyte*) and multicellular diploid (*sporophyte*) stages. In the life cycle, the haploid zoospores (stage 1) divide mitotically to produce haploid multicellular thalluses (stage 2). These may reproduce asexually by zoospores or sexually by gametes (stage 3). Fusion of two gametes produces diploid zygotes (stage 4), which divide mitotically to produce diploid multicellular thalluses (stage 5). Certain reproductive cells of the diploid thallus divide by meiosis to form haploid zoospores, which begin a new cycle. This type of life cycle exhibits *alternation of generations* in that a haploid multicellular phase alternates with a diploid multicellular phase.

Multicellularity in plants arose first in the gametophyte and many green algae have no sporophyte stage, but *Ulva* shows a more advanced life cycle in that both sporophyte and gametophyte stages are present and equally important.

The *Phaeophyta* (brown algae) are multicellular and almost exclusively marine. The thallus may be a filament or a large, complex, three-dimensional structure. The Phaeophyta possess chlorophyll *a* and *c* and the brownish pigment *fucoxanthin*. Reproduction may be asexual or sexual. The life cycle usually exhibits alternation of generations. In many forms the sporophyte and gametophyte are essentially equal, but in other forms the sporophyte is the dominant stage. In a few forms the gametophyte stage is completely absent.

Although the Phaeophyta are thallophytes, they do show some tissue differentiation. They are complex plants that have convergently evolved many similarities to the vascular plants.

The *Rhodophyta* (red algae) are mostly marine seaweeds. Most are multicellular. The Rhodophyta possess chlorophylls *a* and *d*, phycoerythrins, and phycocyanins. The accessory pigments are important in absorbing light and transferring the energy to chlorophyll *a* for photosynthesis. The life cycles of the red algae are complex; alternation of generations is usual. Flagellated cells never occur.

The movement onto land The Embryophyta have evolved numerous adaptations for life on land. Life probably arose in the water, and the evolutionary move to land was not a simple one. A terrestrial environment poses many problems for plants, including obtaining enough water, transporting water and dissolved materials from one part of the plant to another, preventing excessive evaporation, maintaining a moist surface for gas exchange, supporting the plant against gravity, carrying out reproduction, and withstanding extreme environmental fluctuations.

Much of the evolution of the embryophyte plants can best be understood in terms of adaptations to help solve these problems. The sterile jacket cells around the multicellular sex organs help protect the enclosed gametes from desiccation. Such male and female sex organs are known as *antheridia* and *archegonia*. All embryophytes are oogamous, and fertilization and early embryonic development take place within the moist environment of the archegonia. The aerial surfaces of the plant are often covered by a waxy cuticle, which prevents excessive water loss.

These plants are biochemically similar to the green algae; they possess chlorophylls *a* and *b*, and the reserve material is starch.

Bryophyta The *Bryophyta* (liverworts, hornworts, and mosses) are small plants that grow only in moist places. Moisture must be available for the flagellated sperm to swim to the egg. Most bryophytes lack vascular tissues for transport and support. The bryophytes show alternation of generations; the haploid stage is dominant. The sporophyte is attached to the gametophyte; it consists of a foot embedded in the gametophyte, a stalk, and a capsule (sporangium). Meiosis occurs within the sporangium, producing haploid spores that fall to the ground, germinate, and eventually develop into mature gametophytes.

Tracheophyta All members of the division *Tracheophyta* possess four important attributes lacking in the algae: a protective layer of sterile jacket cells around the reproductive organs; multicellular embryos retained within the archegonia; cuticles on aerial parts, and xylem. All four are fundamental adaptations for a terrestrial existence.

The *Psilopsida* are the most primitive of the vascular plants. They lack leaves and have no true roots, although they do have underground stems with rhizoids. Photosynthesis occurs in the green, branching stems, some of which bear sporangia at their tips. The life cycle shows an alternation of generations.

The *Lycopsida* (club mosses) show many advances over the Psilopsida. They have true roots, stems, and leaves. Some of the leaves have become specialized for reproduction and bear sporangia on their surfaces; these leaves are called *sporophylls*. In many Lycopsida the sporophylls are compacted together to form a club-shaped structure (strobilus). Some lycopsids produce only one kind of spore (i.e. they are *homosporous*); these develop into a gametophyte that bears both antheridia and archegonia. Others produce two types of spores, large *megaspores* and smaller *microspores*; these plants are *heterosporous*. The megaspores develop into female gametophytes, the microspores into male gametophytes.

Like the lycopsids, the *Sphenopsida* (horsetails) have true roots, stems, and leaves. Some ancient sphenopsids were large trees; much of today's coal was formed from these plants.

The *Pteropsida* (ferns) probably evolved from the Psilopsida. The ferns are fairly advanced plants with a well developed vascular system. The large, leafy fern plant is the diploid sporophyte phase. Most ferns are homosporous; the spores are produced in sporangia on the fertile leaves (sporophylls). Upon germination the spores develop into small, independent, nonvascular gametophytes bearing antheridia and archegonia. The flagellated sperm must swim to the archegonium. The gametophytes therefore require a moist habitat; thus ferns are restricted to a moist environment.

The dominant and best adapted land plants are those belonging to the *Spermopsida* (seed plants). The gametophyte stage of these plants has been much reduced, and the sperm are not free-swimming, flagellated cells. In addition, the embryo is protected from desiccation by a seed coat. Heterospory is characteristic of all seed plants. There are six classes of Spermopsida: five of these are often referred to as the gymnosperms, and the sixth class is the Angiospermae.

The *gymnosperms* appear to be older and less specialized than the angiosperms. The best-known group of gymnosperms is the conifers (class Coniferae), or evergreens. The leaves of most of these plants are small needles or scales.

The pine provides a good example of the seed method of reproduction. The pine tree is the diploid sporophyte stage. It bears two kinds of *cones*: large female cones, which produce spores that develop into female gametophytes; and small male cones, which produce spores that develop into *pollen grains* (the male gametophyte). The pollen grains are released into the air; some may land on the scales of a female cone. The pollen grain develops a *pollen tube*, which grows through the tissues of the female sporangium into one of the archegonia. There it discharges its two sperm nuclei, one of which fertilizes the egg. The zygote develops into an embryo, and a tough seed coat surrounds it and its stored food material. Finally, the *seed* is shed from the cone.

The Angiospermae are the dominant land plants. Their reproductive structures are the flowers. In a flower the outer leaflike *sepals* protect the inner floral parts; the *petals* often attract animal pollinators; the *stamens* (filament and anther) are the male reproductive organs; and the central *pistil* is the female reproductive organ. Each pistil consists of a *stigma*, *style*, and *ovary*. Within the ovary are one or more sporangia called *ovules*. Meiosis occurs in each ovule, producing one functional megaspore, which divides to produce (in most species) a haploid, seven-celled, eight-nucleate, female gametophyte.

Each anther has sporangia in which meiosis occurs, producing haploid microspores. These develop into thick-walled, two-nucleate pollen grains—the male gametophytes. A pollen grain germinates when it lands on the pistil; a pollen tube grows through the pistil, enters the ovary, and discharges two sperm nuclei into the female gametophyte. "Double fertilization" occurs: one sperm nucleus fertilizes the egg, the other combines with the two polar nuclei to form a triploid nucleus, which will give rise to the *endosperm*.

After fertilization the ovule matures into a

seed, which consists of a seed coat, stored food (endosperm), and embryo. The seeds are enclosed in fruits that develop from the ovaries and associated structures.

Angiosperms differ from gymnosperms in possessing xylem vessels and in many reproductive characteristics: angiosperms have greatly reduced gametophytes; their pollen tubes have farther to grow; they have "double fertilization"; their endosperm is triploid; and the seeds are enclosed within fruits.

For these reasons the angiosperms are the most successful and dominant land plants. They are divided into two subclasses: the Monocotyledonae (embryo has one cotyledon) and the Dicotyledonae (embryo has two cotyledons). There are many basic differences between the two groups.

QUESTIONS

Testing recall

Match each characteristic listed below with the division with which it is most closely associated.

a Chlorophyta
b Phaeophyta
c Rhodophyta

1 They contain many unicellular and multicellular forms.

2 They possess the pigment fucoxanthin.

3 They possess the pigments chlorophyll *a* and *b*.

4 They possess the pigments phycoerythrin and phycocyanin.

5 Some forms show a primitive type of tissue differentiation.

6 Many forms show alternation of generations with a dominant sporophyte.

7 Accessory pigments enable these plants to survive in deep water.

8 Land plants probably evolved from this division.

9 They lack flagellated cells in life cycle.

10 They store starch as reserve material.

Mark each characteristic below T if it is found in the Thallophyta, and E if it is found in the Embryophyta. (Some characteristics may be found in both groups.)

11 tissue differentiation

12 isogamy

13 multicellular sex organs

14 sterile jacket cells around reproductive organs

15 dominant gametophyte

16 zygote development outside female reproductive organs

17 cuticle on aerial parts

18 embryo development within archegonium

The table below lists some of the problems faced by land plants. Complete the chart by filling in the features the land plants evolved to solve these problems. The completed chart will be a useful study aid.

Problem	Bryophyta	Psilopsida	Lycopsida	Sphenopsida	Pteropsida	Spermopsida
Obtaining enough water						
Transport of materials						
Supporting plant against the pull of gravity						
Preventing excessive water loss						
Carrying out reproduction with little water						
Protection for the embryo						

Testing knowledge and understanding

Choose the one best answer. (Questions 2–7 include groups covered in other chapters.)

1 Which of the following best describes the type of sexual reproduction in which the gametes are all motile but are of two sizes, one type big and one type small?

   a homospory   *d* oogamy
   b heterospory   *e* isogamy
   c anisogamy

2 The evolutionary trend toward multicellularity is best demonstrated by comparing various members of the

   a Phaeophyta.   *d* Chlorophyta.
   b Bryophyta.   *e* Rhodophyta.
   c Cyanophyta.

3 A certain plant is photosynthetic, does not use true starch as its principal carbohydrate storage product, has a rudimentary phloem-like tissue but no xylem, and possesses multicellular sex organs not enclosed by sterile jacket cells. To which group does the plant probably belong?

   a Bryophyta   *d* Chlorophyta
   b Tracheophyta *e* Phaeophyta
   c Rhodophyta

4. In which one of the following divisions would you *least* expect to find a multicellular sporophyte stage in the life cycle?

   a Euglenophyta
   b Chlorophyta
   c Phaeophyta
   d Rhodophyta
   e Bryophyta

5. In the ocean, which algae would normally be found in the deepest water?

   a Chlorophyta
   b Chrysophyta
   c Phaeophyta
   d Rhodophyta

6. The algal division that has evolved the most complex multicellular thallus is

   a Cyanophyta.
   b Chlorophyta.
   c Chrysophyta.
   d Phaeophyta.
   e Rhodophyta.

7. "Higher" plants have probably evolved from which algal line?

   a Chlorophyta
   b Rhodophyta
   c Phaeophyta
   d Chrysophyta
   e Pyrrophyta

8. Which of the following does *not* function specifically as an adaptation to the land environment?

   a multicellular sex organ with jacket cells
   b structures to attach the plant to the substrate
   c cuticle
   d embryonic development within the archegonium
   e vascular tissue

9. The movement of plants from water to land and their subsequent increase to large size was made possible by the appearance of

   a collenchyma.
   b parenchyma.
   c xylem and phloem.
   d epidermis.
   e cortex.

10. As plants occupied drier habitats they could no longer rely on the transport of sperm in water. The gymnosperms first overcame this problem by evolving

    a pollen.
    b cones.
    c motile sperms.
    d eggs and sperms in same plant.
    e mature sperms produced only in the rainy season.

11. Along with the seed, the seed plants have evolved several additional adaptations to the land environment. Which of the following is *not* such an adaptation?

    a Flagellated gametes are not required for seed formation.
    b The female gametophyte is protected from desiccation by the surrounding tissues of the sporophyte.
    c The seed and/or associated structures serve as a means of dispersal.
    d The method of seed formation introduces a new type of genetic recombination.
    e The seed has its own food supply for the enclosed embryo.

12. You are given an unknown plant to study in the laboratory. You find that it has chlorophyll, but no xylem. Its multicellular sex organs are enclosed in a layer of jacket cells. Its gametophyte stage is free-living. The plant probably belongs in the

    a Chlorophyta.
    b Phaeophyta.
    c Bryophyta.
    d Rhodophyta.
    e Tracheophyta.

13. The earliest tracheophytes, which probably gave rise to the other vascular plants, belong to the subdivision

    a Spermopsida.
    b Lycopsida.
    c Psilopsida.
    d Pteropsida.

14. Which one of the following statements is true?

    a Gymnosperms have a free-living gametophyte stage in their life cycle.
    b Most gymnosperms have flagellated sperm cells.
    c The female gametophyte of a flowering plant consists of four haploid megaspores.
    d The cells in a maple leaf are haploid while those of a moss are diploid.
    e The endosperm of a spruce seed is haploid.

15. In angiosperms, the sporophylls are the

    a stamens and pistils.
    b sepals and petals.
    c sepals and stamens.
    d anthers and petals.
    e sepals and pistils.

16. A flower has six petals, three sepals, nine stamens, three ovules, and one ovary. How many seeds will this flower produce?

    a one
    b three
    c six
    d nine
    e 27

17. The plant in question 16 is a

    a monocot.
    b dicot.

In questions 18-22, match the specific part of the flower with its function. Answers may be used more than once or not at all.

a anther
b ovary
c ovule
d sepal
e stigma

18 Contains the female gametophyte.

19 Male gametophyte is formed here.

20 Matures to form the fruit.

21 Receives the pollen grain.

22 Forms the seed.

23 The endosperm of a flowering plant is triploid because

   a it results from the fusion of a normal gamete and a diploid gamete resulting from nondisjunction.
   b two male nuclei fuse with one female nucleus.
   c two polar nuclei fuse with a sperm nucleus.
   d none of the above.

For further thought

1 Draw an evolutionary tree for the kingdom Plantae.

2 The Phaeophyta are complex plants that have convergently evolved many similarities to the vascular plants. Many live in the intertidal zone, where they are subject to wave action and the danger of desiccation when the tide is out. What adaptations have they evolved to survive in such a habitat?

3 As tracheophytes evolved, the sporophyte generation became more dominant, and the gametophyte smaller and more dependent on the sporophyte. What is the adaptive advantage of the sporophyte stage over the gametophyte?

4 What is the relationship between the gametophyte and sporophyte of a liverwort? a fern? a pine tree? a flowering plant?

5 Discuss some of the adaptations by means of which cross-pollination is encouraged in the flowering plants.

6 Mendel, in his experiments with the garden pea, found that peas are normally self-fertilized since maturation of the floral organs occurs before the flower has opened. What procedure would you follow to bring about cross-fertilization in such a plant?

Chapter 24

THE FUNGAL KINGDOM

KEY CONCEPTS

1 The fungi are a diverse group of absorptive heterotrophic, but sedentary, organisms that have great economic importance and play a vital role as decomposers.

2 The eucaryotic cells of fungi are walled and are usually organized into threadlike hyphae in which the cellular partitions are often incomplete.

OBJECTIVES

After studying this chapter and reflecting on it, you should be able to

1 Give the main characteristics of the fungi.

2 Compare the four phyla with respect to: occurrence, presence or absence of cellular partitions in their hyphae, mode of reproduction, and economic importance.

3 Give four ways in which the Oomycota differ from the other fungi.

4 Discuss the life cycle of the black bread mold *Rhizopus*.

5 Distinguish among conidium, ascus, and basidium.

6 Diagram the life cycle of an ascomycete and a basidiomycete.

SUMMARY

The fungi are saprophytic or parasitic organisms that are primarily multicellular or multinucleate. The partitions between the cells are generally either absent or partial, so the cytoplasm is continuous. The cell wall usually contains *chitin*. The cells are usually organized into branched filaments called *hyphae*, which form a mass called a *mycelium*. Reproduction may be sexual or asexual, but the haploid stages are usually dominant. The fungi are important decomposers and many are of great economic importance. Four phyla of true fungi are recognized: Oomycota, Zygomycota, Ascomycota, and Basidiomycota. Characteristics related to sexual reproduction are important in defining the groups.

The *Oomycota* (water molds) are generally aquatic organisms that differ markedly from the other phyla in that they have flagellated zoospores, are oogamous and diploid, and lack chitin in their cell walls. The hyphae lack cellular partitions (septa).

Members of the *Zygomycota* (conjugation fungi) are widespread as saprophytes in soil and dung. Their hyphae characteristically lack cross-walls, so the multinuclear cytoplasm is continuous (coenocytic). Reproduction may be asexual (by spores) or sexual. Sexual reproduction occurs by conjugation between two cells from hyphae of two different mycelia. These gamete cells fuse to form a zygote, which develops a thick wall and becomes dominant. At germination, the nucleus undergoes meiosis and a hypha develops. It releases asexual spores that grow into new mycelia. *Rhizopus* (black bread mold) is an example of this phylum.

The *Ascomycota* (sac fungi) are a diverse group, varying from unicellular yeasts through powdery mildews to cup fungi. The hyphae are septate, but the septa usually are incomplete and have large holes. All produce a reproductive structure called an *ascus* during their sexual cycle. An ascus is a sac within which haploid spores, usually eight, are produced. These fungi also reproduce asexually by means of special spores called *conidia*. The fungal member of a *lichen* is generally an Ascomycota.

The *Basidiomycota* (club fungi) include the puffballs, mushrooms, toadstools, and bracket fungi. The large fruiting bodies of these fungi are composed of compacted, septate hyphae. During sexual reproduction, hyphae from two different mycelia with uninucleate cells unite and give rise to hyphae with binucleate cells. Certain terminal cells become zygotes when their two nuclei fuse. The zygote then becomes a *basidium*, a club-shaped reproductive structure. Meiosis occurs within the basidium, and four haploid spores are produced. Each spore may give rise to a new mycelium.

QUESTIONS

Testing recall

Complete the following chart.

	Hyphae (septate or non-septate)	Flagellated Cells (yes or no)	Dominant Stage (haploid or diploid)	Sexual Reproductive Structure	Representative Members	Economic Importance
Oomycota						
Zygomycota						
Ascomycota						
Basidiomycota						

Testing knowledge and understanding

Choose the one best answer.

1. In most of the fungi, the dominant stage is
   - a haploid.
   - b diploid.

2. Which one of the following pairs is incorrectly matched?
   - a Oomycota–diploid
   - b Zygomycota–*Rhizopus*
   - c Zygomycota–septate hyphae
   - d Ascomycota–lichens
   - e Ascomycota–conidia

3. During their sexual cycle, the Zygomycota produce a(n)
   - a conjugation tube.
   - b ascus.
   - c basidium.
   - d flagellated gamete.
   - e bud.

4. The fungi that produce a diploid resting spore and have haploid hyphae belong to the
   - a Ascomycota.
   - b Oomycota.
   - c Zygomycota.
   - d Basidiomycota.

5. Which one of the following is *not* found in the Basidiomycota?
   - a a mycelium
   - b a basidium
   - c conidia
   - d binucleate hyphae.
   - e septate hyphae.

6. When we eat a mushroom we are eating
   - a a mycelium.
   - b the fruiting body.
   - c basidia.
   - d hyphae.
   - e all of the above.

7. A student was given an unknown organism to identify. Although it looked like a solid mass of tissue, the student found it was made up of a mass of filaments, the cells of which lacked chlorophyll. Cross walls were present in the filaments, however, and two nuclei were found in many of the cells. The reproductive structures were club-shaped and four spores were found on the tip of these structures. The student decided that this plant probably belonged in the
   - a Schizomycetes.
   - b Ascomycota.
   - c Oomycota.
   - d Basidiomycota.
   - e Zygomycota.

Chapter 25

THE ANIMAL KINGDOM

KEY CONCEPTS

1 Characteristics such as level of organization, type of symmetry, segmentation, embryonic development, and larval forms have been used to establish hypothetical phylogenic relationships among the various animal groups.

2 The Porifera differ greatly from all other multicellular animals and probably arose independently.

3 The two radiate phyla—Coelenterata and Ctenophora—comprise radially symmetrical animals whose bodies are at a relatively simple level of construction.

4 The Platyhelminthes and Nemertina are thought to be the most primitive bilaterally symmetrical animals; they are composed of three well-developed germ layers and are acoelomate.

5 A major split probably occurred in the animal kingdom soon after the origin of a bilateral organism. One evolutionary line led to the phyla in which the blastopore becomes the mouth, the *Protostomia*; the other line, called the *Deuterostomia*, led to the phyla in which the blastopore becomes the anus and a new mouth is formed.

6 In several protostome phyla (Acanthocephala, Ectoprocta, Aschelminthes) the body cavity is not completely enclosed by mesoderm; it is a pseudocoelom. All the other protostome phyla have true coeloms (at least in some stage of their development); in most groups these arise as a split in the initially solid mass of mesoderm.

7 The arthropods are generally regarded as the most highly evolved representatives of the protostome line.

8 The Deuterostomia (Echinodermata, Hemichordata, and Chordata) differ from the Protostomia in the formation of the anus from the blastopore, in having radial and indeterminate cleavages, in the origin of the mesoderm as pouches, and in the formation of the coelom.

9 The chordates and echinoderms are believed to have evolved from a common ancestor at some remote time.

10 All chordates possess, at some time in their life cycle, a notochord, pharyngeal gill pouches, and a dorsal hollow nerve cord.

11 The vertebrates may be regarded as the most highly evolved representatives of the deuterostome line.

OBJECTIVES

After studying this chapter and reflecting on it, you should be able to

1 Give three characteristic features of the Porifera, and give reasons why the Porifera are believed to have evolved independently of other multicellular animals.

2 Give the distinguishing features of the Coelenterata and Ctenophora.

3 Name and give the characteristics of each of the three coelenterate classes, and name a representative organism of each class. Describe their life cycles.

4 Contrast Haeckel's gastraea hypothesis, the plakula hypothesis, and the ciliate hypothesis of the origin of the Metazoa.

5 Distinguish between the terms acoelomate, pseudocoelomate, and coelomate.

6 Give the major characteristics of the Platyhelminthes, and list and distinguish among the three classes.

7 Discuss the tapeworm's special adaptations for parasitism.

8 Give two important evolutionary advances of the Nemertina.

9 Give four fundamental differences between the Protostomia and Deuterostomia.

10 Give the distinguishing features of the Aschelminthes, and describe the structure and distribution of the Nematoda. Name four nematodes that parasitize human beings.

11 Describe and give the function of the lophophore, and name three phyla possessing this structure.

12 Describe the body plan common to all molluscs. Name and describe the six classes, and give representative examples of each.

13 Describe ways in which the cephalopods have convergently evolved many similarities to vertebrates.

14 List the classes and identifying characteristics of the annelids, and give representative examples of each class.

15 Describe the ways in which the Onychophora resemble the annelids and ways in which they resemble the arthropods.

16 Give the distinguishing features of the arthropods; list and describe the three subphyla, and give representative examples of each.

17 Differentiate between chelicerae and mandibles.

18 Give the distinguishing characteristics of the Crustacea, Chilopoda, Diplopoda, and Insecta, and list representative examples of each.

19 Briefly describe the structure of the generalized insect body.

20 List the phyla of the Deuterostomia.

21 List the classes and the identifying features of the Echinodermata, and give representative examples of each class.

22 Discuss and give evidence for the echinoderm theory of chordate origin.

23 Give three distinguishing features of the Chordata, and list the three subphyla.

24 Describe the adult and larval forms of the tunicates, and suggest an evolutionary relationship between the tunicates and vertebrates.

25 List the vertebrate classes and the general characteristics of each.

26 Trace the evolutionary relationships between the major classes of vertebrates.

27 List eight characteristics shared by all primates.

28 Discuss current ideas of the evolutionary history of man.

SUMMARY

The animal kingdom includes many diverse groups of organisms.

The *Porifera* (sponges) differ greatly from other multicellular animals and are believed to have evolved on a line of their own. They are multicellular animals at a primitive, pre-tissue level of organization. Sponges are filter feeders; their bodies are perforated sacs through which water flows. The flagellated collar cells lining the interior are remarkably similar to protozoan collared flagellates.

The radiate phyla These phyla—Coelenterata and Ctenophora—comprise radially symmetrical animals whose bodies have definite tissue layers but no distinct internal organs. They have a gastrovascular cavity and lack a coelom. Their bodies consist of two well-developed tissue layers, an outer epidermis (ectoderm) and an inner gastrodermis (endoderm), with a third layer of mesoglea (variable in its development) between them.

The *Coelenterata* include the hydra, jellyfishes, sea anemones, and corals. There is some division of labor among their tissues, but it is never as complete as in most bilateral multicellular animals, and most functions performed by mesodermal tissues in other animals are performed by ectodermal or endodermal cells in coelenterates. There are three classes: Hydrozoa, Scyphozoa, and Anthozoa.

Many hydrozoans have a sedentary *polyp* stage that alternates with a free-swimming jellyfishlike *medusa* stage. Certain cells of the medusa produce gametes. The zygote develops into an elongate ciliated *planula* larva, which eventually settles and gives rise to the polyp. In the scyphozoans (jellyfishes), the medusa stage is dominant; in the anthozoans (sea anemones and corals), it is absent altogether. The Anthozoa are the most advanced of the Coelenterata.

The *Ctenophora* differ from the coelenterates in having mesodermal muscles and in possessing eight rows of ciliary plates; they lack the coelenterate's nematocysts and polymorphic life cycle.

Origin of the Metazoa Many biologists hypothesize that the *Metazoa* evolved from a hollow sphere colonial flagellate (*blastaea*) with an intermediate *planuloid* stage. Others have suggested that the Metazoa evolved from a two-layered, flattened, creeping *plakula*. Its ventral surface may have been specialized for nutrition; a temporary digestive system could have been formed by elevating part of its body. Support for this hypothesis has come from the rediscovery of the organism *Trichoplax*, which bears a marked resemblance to the hypothetical plakula. Still other biologists have suggested that the multicellular animals arose from multinucleate ciliates in which cellular membranes formed around each nucleus.

Acoelomate Bilateria The *Platyhelminthes* and *Nemertina* are regarded as the most primitive bilaterally symmetrical animals. Their bodies are composed of three well-developed tissue layers; there is no coelom.

The *Platyhelminthes* (flatworms) are dorsoventrally flattened, elongate animals with (usually) a gastrovascular cavity. The flatworms have advanced to the organ level of construction. The more extensive development of mesoderm probably made this advance possible. The phylum is divided into three classes: Turbellaria, Trematoda, and Cestoda.

The *Turbellaria*, of which planarians are examples, are free living. Since one order, the Acoela, is very primitive, some biologists hypothesize that an *acoeloid* organism could have arisen from a planuloid ancestor, or from a multinucleate ciliate, and that the more complex flatworms and other metazoan phyla then evolved from such an acoeloid organism.

The Trematoda (flukes) and Cestoda (tapeworms) are entirely parasitic. They have evolved many adaptations for parasitism: a resistant cuticle, loss or reduction of structures, extremely well-developed reproductive systems, and elaborate life cycles.

The members of the phylum *Nemertina* probably evolved from the turbellarian flatworms. However, they have a *complete digestive system* and a simple blood circulatory system.

Divergence of the Protostomia and Deuterostomia The fate of the embryonic blastopore differs in various animal groups. In many animals the blastopore becomes the mouth, and a new anus develops; in others the situation is reversed. This fundamental difference in development suggests that a major evolutionary split occurred soon after the origin of a bilateral ancestor. One evolutionary line led to all the phyla in which the blastopore becomes the mouth, the *Protostomia* (first mouth). The other line led to the phyla in which the blastopore forms the anus and a new mouth is formed, the *Deuterostomia* (second mouth). The protostomes and deuterostomes also differ in the determinateness and pattern of the initial cleavages, the mode of origin of the mesoderm and coelom, and the type of larva.

The pseudocoelomate Protostomia A true coelom is a cavity enclosed entirely by mesoderm and located between the digestive tract and body wall. In several protostome phyla the body cavity is partly bounded by ectoderm and endoderm; it is called a *pseudocoelom*.

The largest and most important pseudocoelomate phylum is the Aschelminthes. These are small, wormlike animals without a definite head that have a complete digestive tract. The members of two classes are very abundant: the Rotifera (wheel animals) and the Nematoda (roundworms). Many nematodes are parasitic in plants and animals. *Trichinella*, *Ascaris*, hookworms, pinworms, and filaria worms parasitize humans.

Coelomate Protostomia Sometime during their development, all the other protostome phyla possess true coeloms, which usually develop as a split in the mesoderm. All have a complete digestive tract, and most have well-developed circulatory, excretory, and nervous systems.

Three small phyla, *Phoronida*, *Ectoprocta*, and *Brachiopoda*, resemble one another in having a feeding device called a *lophophore*—a fold that encircles the mouth and bears numerous ciliated tentacles.

The phylum *Mollusca* is the second largest in the animal kingdom; it includes the snails and slugs, clams and oysters, and squids and octopuses. The molluscs are soft-bodied animals composed of a ventral muscular *foot*, a *visceral mass*, and a *mantle*, which covers the visceral mass and usually contains glands that secrete a shell. The mantle often overhangs the visceral mass, thus enclosing a *mantle cavity*, which frequently contains gills.

Molluscs have an open circulatory system. Most marine molluscs pass through one or more free-swimming larval stages, but the freshwater and land snails complete the corresponding developmental stages in the egg. The molluscs are usually divided into six classes: Amphineura (chitins), Monoplacophora, Gastropoda (snails), Scaphopoda (tusk shells), Pelecypoda (bivalves), and Cephalopoda (squids, octopuses). The cephalopods are specialized for rapid movement and predation; they have convergently evolved many similarities to vertebrates.

The phylum *Annelida* (segmented worms) is usually divided into three classes: Polychaeta, Oligochaeta, and Hirudinea. The polychaetes are marine animals with a well-defined head bearing eyes and antennae. Each body segment bears a pair of *parapodia*, which function in movement and gas exchange. The Oligochaeta include earthworms and freshwater species; they lack a well-developed head and parapodia, and have few setae. The Hirudinea are the leeches. They are the most specialized of the annelids and show little internal segmentation.

The members of the phylum *Onychophora* are of special interest because they combine annelid and arthropod characters and are regarded as an early evolutionary offshoot from the line leading to the arthropods from an ancient annelidlike ancestor.

The phylum *Arthropoda* is the largest of the phyla. Arthropods are characterized by a jointed chitinous exoskeleton and jointed legs. They have elaborate musculature, a well-developed nervous system and sense organs, and an open circulatory system.

Arthropods are believed to have evolved from a polychaete annelid or from the ancestor of the polychaetes. The arthropod body may be an elaboration and specialization of the annelid's segmented body. Evidence indicates that the first arthropods had long wormlike bodies composed of nearly identical segments, each bearing a pair of legs. Among evolutionary trends seen in the arthropods are reduction in the number of segments; groupings of segments into body regions; increasing cephalization; and specialization of some legs for functions other than movement, with loss of legs from other segments.

The Arthropoda can be divided into three subphyla: Trilobita, Chelicerata, and Mandibulata. The *Trilobita* are now extinct; they resembled the hypothetical arthropod ancestor. Fossils show that every segment bore a pair of nearly identical legs. In both the Chelicerata and Mandibulata the tendency towards specialization of some appendages and loss of others is apparent.

The *Chelicerata* include the horseshoe crabs and arachnids (spiders, ticks, mites, scorpions). The body is usually divided into a cephalothorax and an abdomen. The first pair of postoral legs is modified as mouthparts called *chelicerae*. There are five other pairs of appendages on the cephalothorax; in some the first pair is modified as feeding devices called *pedipalps*.

The members of the *Mandibulata* have *mandibles* as their first pair of mouthparts, and two additional pairs of mouthparts called *maxillae*. The Mandibulata include the classes Crustacea (lobsters, crabs, shrimps, water fleas, sow bugs), Chilopoda (centipedes), Diplopoda (millipedes), and Insecta.

The insects are an enormous group of diverse organisms; there are more species of insects than of all other animal groups combined. The insect body is divided into a head, thorax, and abdomen. The head bears many sensory receptors, usually including compound eyes; one pair of antennae; and three mouthparts (mandibles, maxillae, and *labium*) derived from ancestral legs. The upper lip, or *labrum*, may also be derived from ancestral legs. The thorax is composed of three segments, each bearing a pair of legs. In many insects both the second and third segments bear a pair of wings. The abdomen is composed of a variable number (12 or fewer) of segments; the terminal parts may be modified for reproduction.

The Deuterostomia The phyla *Echinodermata*, *Hemichordata*, and *Chordata* are in the Deuterostomia. Members of the phylum *Echinodermata* are exclusively marine, mostly bottom-dwelling animals. The adults are radially symmetrical, but the larvae are bilateral, and it is generally held that the echinoderms evolved from bilateral ancestors. All members have an internal skeleton made of bony plates, a well-developed coelom, and a unique *water-vascular system*. All show pentaradiate symmetry. The echinoderms are divided into five classes: Asteroidea, Ophiuroidea, Echinoidea, Holothuroidea, and Crinoidea.

The *Asteroidea* (sea stars) have a body that consists of a central disc with five arms. The *Ophiuroidea* (brittle stars) have a smaller central disc and five long thin arms. The *Echinoidea* (sea urchins and sand dollars) have no arms, but do have five bands of tube feet, movable spines, and a complex chewing apparatus. *Holothuroidea* (sea cucumbers) have a much reduced endoskeleton and a leathery body. Gas exchange is by a respiratory tree. *Crinoidea* (sea lilies) are often stalked and sessile.

Members of the phylum *Hemichordata* (acorn worms) show affinities with both the echinoderms and chordates. They resemble the chordates in their pharyngeal gill slits and dorsal nerve cord. However, their ciliated larva (a *dipleurula*) is almost identical to that of the echinoderms. The hemichordates' ties to both these groups have helped clarify the relationships between these two major phyla. The chordates and echinoderms are believed to have evolved from a common ancestor at some remote time.

The chart on page 1044 of your text provides a useful summary of some important characteristics of the animal phyla.

The phylum *Chordata* contains both invertebrate and vertebrate members. It is divided into three subphyla: Urochordata, Cephalochordata, and Chordata. At some time in their life cycle, all chordates have a *notochord*, pharyngeal gill slits, and a dorsal hollow nerve cord.

The adults of the *Urochordata* (tunicates) are sessile filter feeders. The larvae, however, are free-swimming, bilaterally symmetrical, tadpolelike organisms with gill slits, notochords, and dorsal hollow nerve cords.

The members of the *Cephalochordata* (lancelets) are small marine organisms with a permanent dorsal hollow nerve cord and notochord. They are filter feeders and show segmentation.

The *Vertebrata* are characterized by an endoskeleton that includes a segmented backbone composed of a series of vertebrae that develop around the embryonic notochord.

The first vertebrates belonged to the class *Agnatha*; they were fishlike animals encased in an armor of bony plates. They lacked jaws and paired fins. The lamprey and hagfishes are the only jawless fishes found today, and are very unlike their ancient ancestors.

A second group of armored fishes, the *Placodermi*, probably arose from the ancient agnaths. The placoderms had hinged jaws and paired fins. The evolution of jaws was an important advance for the vertebrates; it allowed alternative feeding methods to develop. Two classes evolved from the Placodermi: the Chondrichthyes and Osteichthyes.

The modern *Chondrichthyes* (sharks, skates, rays) have a cartilaginous skeleton, whereas the *Osteichthyes* have a bony skeleton. The primitive bony fishes had lungs in addition to gills. Early in their evolution the class split into two divergent groups. One group underwent great radiation, giving rise to most of the bony fishes alive today; the other group, now mostly extinct, includes the present-day lungfishes and lobe-fin fishes. The lobe-fins had lungs and leg-like fleshy paired fins that may have enabled them to crawl. They appear to be ancestral to the amphibians.

The first members of the class *Amphibia* were quite fishlike, but they became a large and diverse group as they exploited the ecological opportunities on land. Most of the amphibians became extinct by the end of the Triassic; the few that survived were the ancestors of the salamanders, apodes, and frogs and toads. Although the amphibians are terrestrial they must return to water to reproduce since they utilize external reproduction, lay fishlike eggs, and have an aquatic larva.

The *Reptilia* are the first truly terrestrial group. They use internal fertilization, lay amniotic shelled eggs, have no larval stage, and have scaly, relatively impermeable skin. In addition, their legs are larger and stronger than those of amphibians, their lungs are better developed, and their heart is almost four-chambered. All the living reptiles (turtles, snakes, lizards, crocodiles, alligators, and the tuatara) except the crocodiles are directly descended from the stem reptiles. One line of the stem reptiles (thecodonts) gave rise to the crocodiles, flying reptiles, and the dinosaurs. All the dinosaurs became extinct except for one group of specialized descendants: the birds. Another line of reptiles (therapsids) gave rise to the mammals.

The *Aves* (birds) have evolved many adaptations for their active way of life. Birds are warm-blooded (homeothermic), their heart is four-chambered, they have feathers for insulation, their bones are light, and they have an extensive system of air sacs.

The *Mammalia* also are homeothermic and have a four-chambered heart. They have a diaphragm, hair for insulation, a one-bone lower jaw, differentiated teeth, and an enlarged neocortex. Embryonic development occurs within the female; the young are born alive and are nourished by milk from the mammary glands. Early in their evolution the mammals split into two groups: the monotremes (egg-laying mammals) and the group that includes the marsupials and placentals. Marsupial embryos remain in the uterus for a short time and then undergo further development attached to a nipple in the mother's abdominal pouch. Placental embryos complete their development in the uterus.

Evolution of the primates Fossil evidence indicates that the members of the mammalian order *Primates* arose from an arboreal stock of small shrewlike insectivores. Important characteristics of the primates include grasping hands (with nails), increased mobility of the shoulder joint (braced by the clavicle), rotational movement of the elbow, good binocular vision, and expansion of the brain. These traits are associated with an arboreal way of life.

The living primates are classified in two suborders: Prosimii and Anthropoidea. The prosimians (lemurs, tarsiers, and others) are more primitive; they are small arboreal animals with apposable first digits. Two lines of Anthropoidea evolved from the Prosimii; one line led to the New World monkeys, and the other, which split, to the Old World monkeys and the apes.

The living great apes (Pongidae) fall into four groups: gibbons, orangutans, gorillas, and chimpanzees. All are large animals with no tail, a large skull and brain, and long arms. All have a tendency to walk semi-erect. The earliest members of the family Hominidae (humans) probably arose from the same pongid stock that produced the gorillas and chimpanzees. Fossils of *Dryopithecus* suggest that the common ancestor of apes and humans had a low rounded cranium, moderate superorbital ridges, and moderate forward projection of face and jaw. In the course of human evolution the jaw became shorter, the skull more balanced on top of the vertebral column, the brain case larger, the nose more prominent, the arms shorter, the feet flatter, and the big toe ceased being apposable. The various fossil humans are intermediate in these characteristics.

The upright posture and bipedal locomotion of human beings may have evolved as adaptations for freeing the hands to carry objects or manipulate tools and weapons. These adaptations would also have made it easier to maintain surveillance.

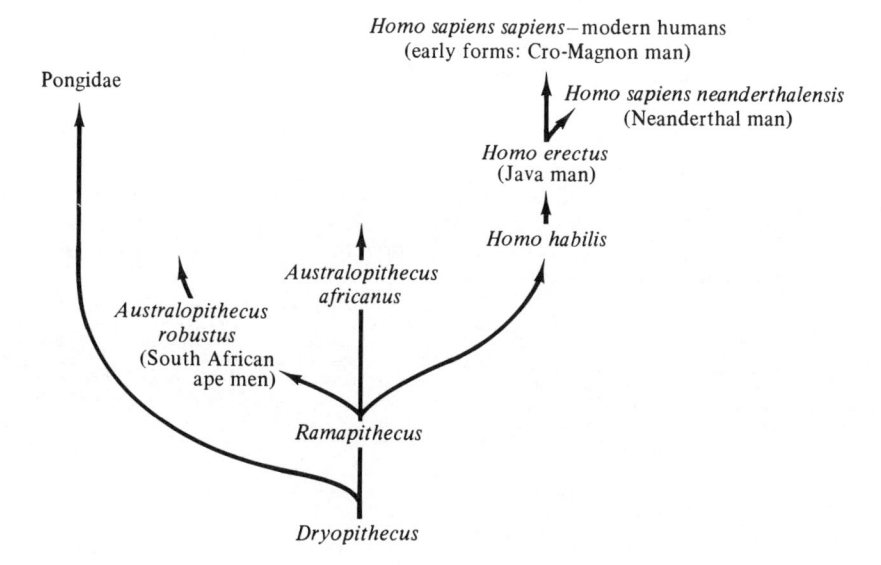

```
                    Homo sapiens sapiens – modern humans
                        (early forms: Cro-Magnon man)
                              ↑
  Pongidae                    ↑ Homo sapiens neanderthalensis
     ↑                          (Neanderthal man)
     ↑                    Homo erectus
     ↑                    (Java man)
     ↑                         ↑
     ↑                    Homo habilis
                   ↑
            Australopithecus
               africanus
       ↑
  Australopithecus
     robustus
  (South African
    ape men)
              Ramapithecus

              Dryopithecus
```

A possible scheme for human evolution is summarized in the figure.

Homo sapiens is a variable species and has geographic variations or races. Many of the differences probably reflect adaptations to different environmental conditions.

Cultural and biological evolution are interwoven, although cultural evolution can proceed more rapidly than biological evolution. Human beings profoundly influence the evolution of other species and now have the ability to influence and control their own evolution.

QUESTIONS

Testing recall

Below is a chart of the animal kingdom, set up as a phylogenetic tree. (It shows only one of many possibilities since much is uncertain about the evolution of animal groups.) A phylogenetic tree can be useful in learning the various groups, particularly if you label the sites of various evolutionary advances. The lines on the tree represent certain evolutionary advances. For example, above line 1, all organisms are multicellular.

Place each of the following characteristics on the appropriate line: complete digestive tract, body cavity which is either a pseudocoelom or a true coelom, true coelom, jointed appendages and exoskeleton, segmentation in adult, three germ layers, and fundamental bilateral symmetry. Also label the Deuterostomia and Protostomia.

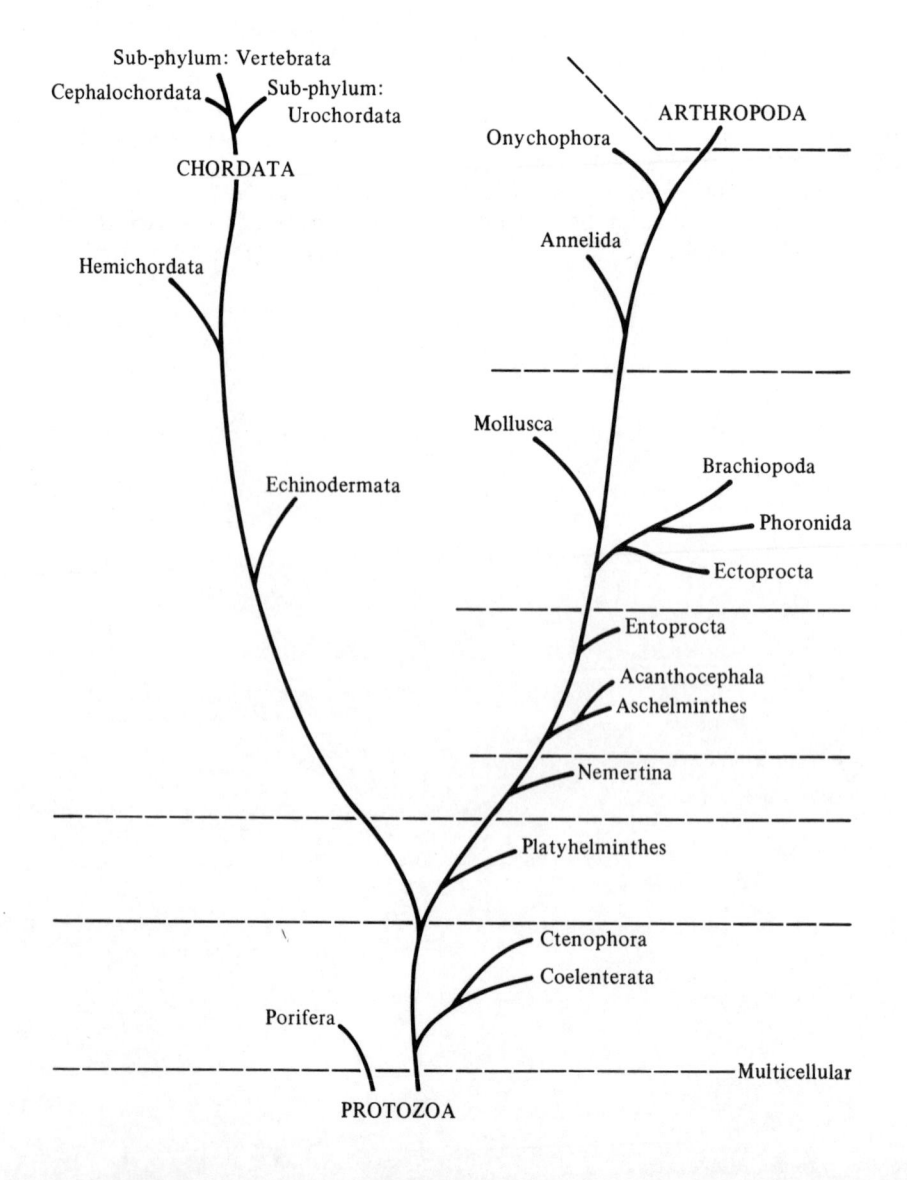

Below are listed some important animal characteristics. For each, write the name(s) of the phylum whose members possess that characteristic.

1. radially symmetrical animals
2. water vascular system
3. jointed legs and exoskeleton
4. lophophore
5. notochord
6. mantle, foot, visceral mass
7. pseudocoelom
8. dipleurula larva
9. bilaterally symmetrical, gastrovascular cavity
10. saclike organisms lined by collar cells

Match each item with the one best answer. Each answer may be used more than once.

a	Agnatha	g	Hemichordata
b	Amphibia	h	Mammalia
c	Aves	i	Osteichthyes
d	Cephalochordata	j	Reptilia
e	Chondrichthyes	k	Urochordata
f	Echinodermata		

11. feathers a characteristic
12. jawless fish with cartilaginous skeletons and scaleless skin
13. saclike marine forms whose larval stage has well-developed notochord and gill slits
14. body hair a characteristic
15. primitive animals in which the adult form possesses pharyngeal gill slits and dorsal nerve cord but no notochord
16. first amniotic vertebrates
17. aquatic animals with bony skeleton and respiration by gills
18. slender elongate animals whose body form closely approaches that of early fishlike vertebrates; have fins, distinct tail, show segmentation, and are filter feeders
19. homeothermic animals with enlarged neocortex
20. marine animals with cartilaginous skeletons believed to have evolved from the ancient Placodermi

Testing knowledge and understanding

Choose the one best answer.

1. The evidence linking the sponges to the flagellated protozoans is

   a the sponge's collar cells.
   b similar photosynthetic pigments.
   c a common reserve food, starch.
   d the volvox-like sponge blastula.
   e There is no such evidence.

2. Which one of the following animals does *not* have a well-developed mesoderm?

   a house fly   d liver fluke
   b jellyfish   e pinworm
   c snail

3. Which one of the following animal groups is probably *least* closely related to all the others?

   a Crustacea   d Polychaeta
   b Arachnida   e Oligochaeta
   c Hemichordata

The following key lists five different animals, each representing a different phylum. The order in which they are listed is random and does not reflect the evolutionary placement of the phyla. Recalling the evolutionary relationships postulated in your text, match the animal whose phylum is the most "primitive" with each characteristic as it appears. (Answers may be used more than once.)

a sponge   d clam
b nematode worm   e planaria
c hydra

4. true coelom
5. circulatory system
6. complete digestive tract
7. well-defined tissue organization
8. bilateral symmetry
9. Which one of the following *arthropod* characteristics was probably the most important adaptation for the land environment?

   a coelom   c exoskeleton
   b segmentation   d circulatory system

10 Which one of the following has a planula larva?

a echinoderms
b mollusca
c arthropods
d coelenterates
e hemichordata

11 Trichinosis, a disease human beings can get from eating insufficiently cooked pork, is caused by a

a virus.
b fungus.
c fluke.
d tapeworm.
e nematode worm.

12 As early vertebrates moved from water to land, loss of water from the body surface became a problem. One "solution" to this problem was the development of

a greater digestive absorptive areas.
b moist skin.
c water-concentrating glands associated with the circulatory system.
d scales.

13 Arthropods and vertebrates may both be regarded, in a sense, as the most highly evolved representatives of their evolutionary lines. They have convergently evolved many striking similarities, presumably in response to similar selection pressures. Which one of the following is *not* a true statement about a similarity between the two groups of animals?

a Most members of both groups possess jointed appendages.
b At least some members of both groups have evolved wings.
c At least some members of both groups have a structurally distinct and highly specialized head region.
d In both groups the circulatory system has become highly specialized for oxygen transport, as is necessary in such active animals.
e Hinged jaws occur in both groups, though their evolutionary derivations are quite different.

14 You are given an unknown species of animal to study in the laboratory. You find that its embryonic blastopore has developed into the adult mouth, that it has a true coelom and a complete digestive tract, that it has an open circulatory system, and that it has a trochophore larval stage. This animal probably belongs in the

a Echinodermata.
b Aschelminthes.
c Mollusca.
d Annelida.
e Nemertina.

15 Which pair has very similar larval stages?

a starfish – vertebrate
b starfish – annelids
c hemichordate – arthropod
d mollusc – starfish
e starfish – hemichordate

16 Which one of the following statements is *not* true of both chordates and echinoderms?

a Both exhibit indeterminate cleavage in early development.
b The embryonic blastopore becomes the anus in both.
c Both have true coeloms.
d Both exhibit radial cleavage.
e Both develop pharyngeal gill pouches and a dorsal hollow nerve cord.

17 According to the classification in your text, the lowest level of classification shared by a frog and a goldfish is the

a kingdom.
b class.
c phylum.
d order.
e subphylum.

18 The jawless fishes

a are now extinct.
b were the direct ancestors of the lobe-fin and lungfish.
c gave rise to the placoderms.
d were very large fish with paired fins and scales much like our modern fish.

19 The lobe-fin fishes appear to have been the direct ancestors of present-day

a freshwater bony fish.
b stem reptiles.
c amphibians.
d ostracoderms.
e placoderms.

20 All members of three vertebrate classes *must* have internal fertilization. These three classes are

a Agnatha, Aves, Mammalia.
b Amphibia, Reptilia, Mammalia.
c Reptilia, Aves, Mammalia.
d Osteichthyes, Amphibia, Reptilia.
e Amphibia, Reptilia, Aves.

21 Which one of the following animals does *not* produce amniotic embryos?

a robin
b alligator
c bullfrog
d human being
e lion

22 A study of fossils suggests that mammals evolved directly from

    a therapsid reptiles.   *d* lobe-fin fishes.
    b dinosaurs.   *e* modern reptiles.
    c Amphibia.

23 A study of fossils suggests that the most immediate ancestors of birds were

    a Amphibia.   *d* mammals.
    b placoderms.   *e* dinosaurs.
    c Osteichthyes.

24 Two fossil vertebrates, each representing a different class, are found in the undisturbed rock layers of a cliff. One is an early amphibian. The other fossil, found in an older rock layer below the amphibian, is most likely to be

    a a dinosaur.
    b a snake.
    c an insectivorous mammal.
    d a fish.
    e a bird.

25 The evolution of which one of the following human characteristics probably *cannot* be explained in part, either directly or indirectly, by the fact that the ancestors of human beings were adapted for an arboreal existence?

    a binocular (3-D) vision
    b freely rotating shoulder joint
    c apposable thumb
    d great reduction of body hair
    e enhanced eye-hand coordination

Questions 26–29. Matching. (Answers may be used more than once or not at all.)

    a Australopithecus
    b Cro-Magnon man
    c Dryopithecus
    d Homo erectus
    e tree-dwelling insectivore

26 earliest representative of our own subspecies

27 *first* true hominid to walk upright

28 common ancestor of apes and man

29 common ancestor of all of the primates

For further thought

1 Discuss the problems faced by animals in terrestrial life, and explain important adaptations in the insects and terrestrial vertebrates that enable them to "solve" these problems.

2 Explain and discuss the pattern of increasing complexity within the animal kingdom (at the phylum level).

3 Arthropods and vertebrates may both be regarded, in a sense, as the most highly evolved representatives of their respective evolutionary lines. They have independently evolved many similar "solutions" to the problems of survival and reproduction. Discuss ways in which the groups resemble each other and ways in which they differ.

ANSWERS

The numbers following the answers refer to pages in the 3rd edition of *Biological Science*.

Chapter 1

1 c,e	13–15	20 e	23	
2 a,b,d	16–19	21 c	23	
3 b,c,d,e	13–18	22 b	22	
4 c,d,e	13–16	23 d	21	
5 a,e	15, 19	24 c	23	
6 d	16–18	25 c	23	
7 b	19	26 c	23	
8 d,f	21	27 a	22	
9 d	21	28 d	17, 18	
10 b	22	29 b	18, 19	
11 d,e	21, 23	30 d	17, 18	
12 g	21	31 a	16, 17	
13 c	23	32 c	17	
14 a	22	33 d	17, 18	
15 b	22	34 d	17, 18	
16 g	21	35 b	18, 19	
17 b	22	36 d	17, 18	
18 f	21	37 c	12	
19 d	21	38 c	7, 8	

Chapter 2

Testing recall

1 false—neutrons 30
2 true 33
3 true 35
4 false—less than 37
5 false—hydrogen bonds 40, 42
6 true 41
7 true 52
8 true 52
9 false—condensation 48
10 true 55
11 true 66
12 true 71
13 a 49
14 c 55
15 b 51, 52
16 b 51, 52
17 a 47
18 b 51, 52
19 d 63, 64

Testing knowledge and understanding

1	b	33	11 b	56, 59, 60
2	b	38	12 a	37, 40, 41, 55
3	b	40, 41	13 b	62
4	a	43	14 e	63–65
5	b	30, 36, 39	15 a	66–70
6	b	30, 36, 46	16 a	70–74
7	d	47, 48, 51	17 a	70
8	c	55	18 b	75
9	c	48	19 c	75
10	b	50		

20 a

$$H_2N-\underset{H}{\overset{H}{C}}-\overset{O}{\overset{\|}{C}}-\underset{H}{\overset{H}{N}}-\underset{CH_2}{\overset{}{C}}-\overset{O}{\overset{\|}{C}}\diagdown_{OH}$$

(with para-hydroxyphenyl group attached to CH₂)

b

$$\begin{array}{c}H-\overset{H}{\underset{|}{C}}-OH\\ H-\overset{|}{\underset{|}{C}}-OH\\ H-\overset{|}{\underset{H}{C}}-OH\end{array} \quad + \quad 3\,C_5H_{11}\overset{O}{\overset{\|}{C}}\diagdown_{OH}$$

c

(disaccharide structure: two pyranose rings linked by O, each with CH₂OH, OH, and H substituents)

Chapter 3

Testing recall

1	e	78	5	g	77
2	h	78	6	c	78
3	a	81	7	b	77
4	d	82	8	f	83

9 *Osmosis* is the movement of water through a semipermeable membrane; *diffusion* is the movement of any particles and may or may not be through a membrane.

10 A *hypertonic* solution has a relatively higher concentration of solute and will gain water by osmosis; a *hypotonic* solution has a relatively lower concentration of solute and will lose water.

11 *Osmotic pressure* is a measure of the tendency of water to diffuse into the system; it varies directly with the concentration of osmotically active particles in the solution, the *osmotic concentration*.

12 In the *unit-membrane model* the proteins form a continuous layer on the inner and outer membrane surfaces; in the *fluid-mosaic model* proteins are found in, on, or through the membrane, but not in continuous layers.

13 *Facilitated diffusion* is passive movement with the concentration gradient; *active transport* is movement against the concentration gradient and requires energy.

14 *Exocytosis* is essentially the reverse of *endocytosis*.

15 *Pinocytosis* and *phagocytosis* differ only in the size of the particles taken in; if the particles are large the process is termed phagocytosis, if liquid or very small, pinocytosis.

16 The *cell wall* is entirely separate from the plasma membrane but the molecules of the *cell coat* attach directly to the molecules of the plasma membrane.

17 The *primary cell wall* is found in all plant cells whereas the thicker *secondary wall* is laid down inside the primary wall in certain plant tissues.

18 *Procaryotic* cells lack almost all of the membranous organelles found in *eucaryotic* cells.

19 *Epithelium* is a type of animal tissue that covers internal and external body surfaces whereas *epidermis* is the outermost layer of cells of a plant or animal body.

20

¹M	I	C	R	O	F	I	L	A	M	E	N	T	²S		
I									M					³V	
T		⁴G		⁵P	E	R	M	E	A	S	E		⁶C	A	
O		O		E					⁷P	O	R	E	L		⁸C
⁹C	H	L	O	R	O	P	L	A	S	T		T	L	U	Y
H		G		O						H		¹⁰N	L	O	T
O		I		X			¹¹L		¹²N	U	C	L	E	U	S
N				I			Y		¹³R		C				O
D				S			S		I		L				L
¹⁴R	N	¹⁵A		O		¹⁶M	I	C	R	O	T	U	B	U	L	E	S
I		C				E			O		O			¹⁷D	
A		T			¹⁸H	Y	P	O			L			N	
		I				M				¹⁹C	I	L	²⁰I	A	
		V											S		
		²¹E	R		²²H	Y	P	E	R						
									²³C	E	N	T	R	I	O	L	E

21 E 120	31 i 124	**Chapter 4**
22 B 120	32 d 122	
23 E 120	33 b 130	**Testing recall**
24 P 120	34 f 122	
25 B 120	35 j 126	1 First 138
26 E 120	36 g 133	2 potential 137
27 E 120	37 k 124	3 reduced 141
28 B 120	38 l 124	4 stores 147
29 B 120	39 h 133	5 red 144
30 E 120	40 e 129	6 ATP, NADP$_{re}$ 150

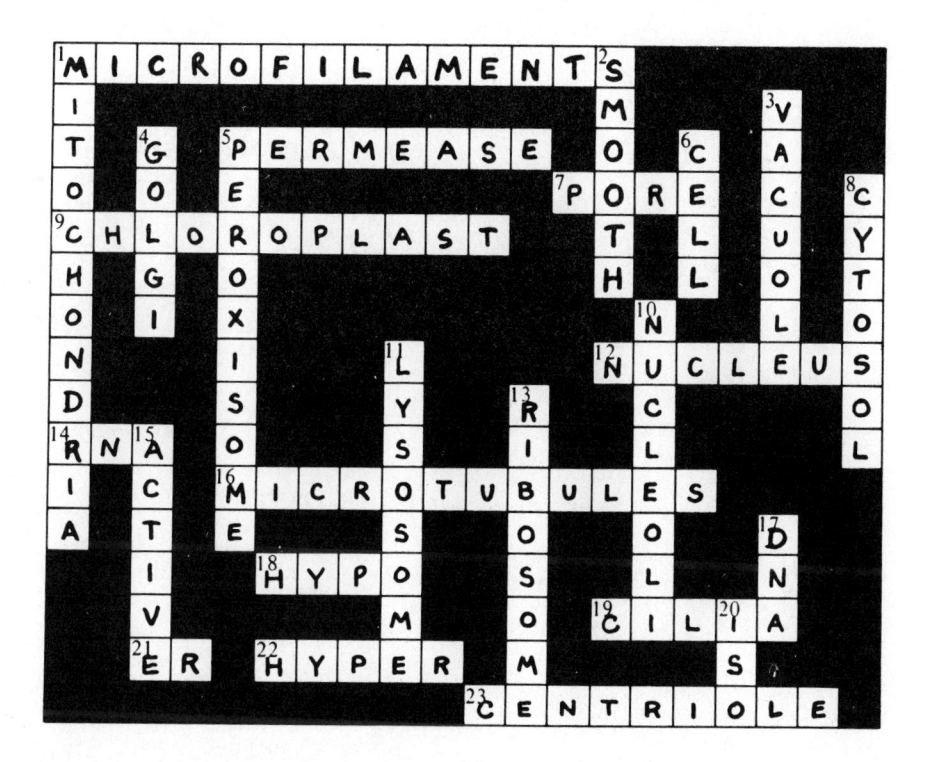

Testing knowledge and understanding

1 b 86	12 e 88	7 destroys 154
2 c 87	13 c 97	8 pyruvic acid 165
3 a 86	14 b 83, 91	9 CO_2, ethyl alcohol 166
4 a 87	15 d 83	10 mitochondria 176
5 c 87	16 a 83	11 electron-transport chain 175
6 b 86	17 d 107, 120	12 poikilothermic 181
7 a 89	18 c 50, 96, 97, 122–133	13 mouse 182
8 d 86	19 d 93	14 d 153
9 a 86	20 a 103, 104	15 e 146, 148, 153
10 a 89	21 c 111	16 c 146, 148
11 d 88	22 c 99, 101, 104, 109	17 b 148 23 a 146
		18 d 153 24 d 164
		19 c 157 25 e 177
		20 d 157 26 a 172
		21 b 149 27 e 167
		22 c 146, 149 28 e 167

Testing knowledge and understanding

1	a	141	10	a	174
2	a	149	11	b	175
3	c	150	12	d	171, 173
4	e	149	13	c	174
5	e	146	14	b	174-175
6	e	147-150, 157	15	c	181
7	d	159-162			
8	b	148-150, 171-172			
9	b	151, 163, 164			

Chapter 5

Testing recall

1	e	197, 199	11	e	232, 224
2	a	197, 199	12	f	229, 224
3	f	195, 199	13	g	231, 224
4	a	197, 199	14	g	233, 224
5	c	197, 199	15	e	232, 224
6	a	197, 199	16	b	235, 224
7	b	197, 199	17	d	230, 224
8	e	197, 199	18	g	225, 224
9	a	223, 224, 229	19	c	229, 224
10	d	224, 230			

Testing knowledge and understanding

1	b	189	16	e	224
2	e	193	17	d	229, 232
3	d	191	18	b	225
4	d	191	19	c	229
5	b	197, 200	20	d	232
6	c	204	21	c	232, 235
7	e	206	22	d	235
8	b	206			
9	d	211, 213, 214, 216			
10	c	202, 203			
11	b	215			
12	d	215, 217			
13	d	222			
14	c	217, 219, 225			
15	b	233			

Chapter 6

Testing recall

1 oxygen 237
2 1 238
3 diffusion, moist 238
4 mesophyll 239
5 open 241
6 the cuticle, closing the stomata 239, 242
7 lenticels 243
8 evaginated, invaginated 245, 248
9 facilitates 246
10 faster 244
11 invaginated, lungs, tracheae 248
12 diffusion, alveoli, capillaries 251
13 nasal cavities 250
14 positive, draw, negative 251, 253
15 decreases 251
16 lungs 253
17 spiracles, tracheae 255
18 large surface area, moist surface, method of transport, protection for fragile surface 239

Testing knowledge and understanding

1	d	239	7	e	250, 251
2	e	242	8	e	251
3	a	238, 239	9	d	238, 250, 253
4	a	245	10	c	252
5	d	246	11	d	255
6	d	246			

Chapter 7

Testing recall

1 false—inside 261
2 true 261, 262
3 false—xylem 262
4 false—both are dead 266
5 true 270
6 false—less active 262, 266, 270
7 true 271-272
8 false—upward and downward 274
9 true 274
10 true 277
11 false—no capillaries 279
12 false—only endothelium 282
13 true 292
14 true 282, 283
15 true 283
16 true 283
17 true 293
18 false—less 292
19 true 291-292, 294
20 false—wettable foreign surface or damaged tissue 298

21	true	304	28	b	280
22	true	304	29	a	281
23	true	301	30	e	280
24	false—HCO_3^-	305	31	d,e	280
25	true	305-306	32	b,e	283
26	b	280	33	d,e	280
27	a	283	34	b,d	282

35 a capillaries of lung, venules, pulmonary vein, left atrium, left ventricle, aorta, arteries leading to brain, arterioles, capillaries of brain, venules, vein in brain   280-281
   b artery in left leg, arterioles, capillaries, venules, veins, posterior vena cava, right atrium, right ventricle, pulmonary arteries, arterioles, capillaries of lungs, venules, pulmonary vein, left atrium, left ventricle, aorta, artery to left arm   280-281

Testing knowledge and understanding

1 c   262, 270, 274
2 a   261, 262
3 c   262, 270, 274
4 e   266
5 c   261-262
6 d   270-272, 277
7 a   272, 277
8 c   271-272
9 d   274-276
10 a   277
11 d   278, 279, 282-283
12 c   282
13 a   285
14 a   285
15 c   280-281
16 c   280-283
17 a   224, 280-281, 311
18 b   280-281
19 c   280-281
20 b   281
21 c   281
22 c   291-292
23 b   285, 287
24 b   294-295
25 e   294-295
26 a   298-299
27 a   302-304
28 c   301-303
29 b   305

Chapter 8

Testing recall

1 equal to, equal to   309
2 equal to, equal to   315
3 less than, equal to   317
4 equal to, greater than   310
5 freshwater invertebrates (d), bony fish (e), amphibians (f)   315-317
6 marine sharks (a), marine invertebrates (b)   315, 317
7 marine bony fish (c), mammals (g), reptiles (h)   317
8 marine sharks (a), marine invertebrates (b), and marine bony fish (c)   317
9 freshwater invertebrates (d), freshwater bony fish (e), amphibians (f)   315-317
10 sharks, fishes, invertebrates   315
11 amphibians (f), mammals (g)   318

12 A glomerulus   323
   B Bowman's capsule   323
   C proximal convoluted tubule   323
   D descending loop of Henle   323
   E ascending loop of Henle   323
   F distil convoluted tubule   323
   G collecting tubule   323
13 a A   323, 326
   b D   327
   c G   327
   d E   326
14 A, B, C, F   323

Testing knowledge and understanding

1 d   310
2 b   319-320
3 b   317
4 a   317
5 b   315-316
6 b   322
7 d   311
8 b   313
9 e   314
10 e   318, 330
11 b   315, 318
12 a   324, 326
13 c   326
14 c   324
15 a   328
16 a   328, 329
17 a   326
18 a   313
19 d   313, 314
20 e   327-328
21 a   326-327
22 c   330-332
23 b   326-328, 330-332

Chapter 9

Testing recall

1 d   347
2 c   346
3 f   345
4 b   341
5 a   347
6 e   350
7 b   344
8 f   345
9 b   339-341
10 a   347
11 i   365-366
12 e   370
13 i   365
14 e,i,k,l   374
15 a   372
16 c   373
17 h   362-364
18 c,h,i   364, 366, 374
19 e   369
20 l   386-387
21 d   371
22 f   371
23 j   361
24 b   375-376
25 a   374-375
26 estrous   390
27 FSH   390
28 estrogen   390
29 LH   391
30 ovulation   391
31 corpus luteum   391
32 estrogen, progesterone   391-392
33 progesterone   392
34 GnRH   392
35 FSH, LH   392
36 progesterone   392
37 menstruation   392
38 FSH   392
39 9-10   391
40 14-15   391
41 4-5   391
42 8-10   395

43 placenta   395
44 chorionic gonadotrophin   395
45 progesterone   395
46 estrogen, progesterone   396
47 estrogen   396
48 oxytocin   396
49 seminiferous tubules, ducts to epididymis, vas deferens, urethra, vagina, uterus, oviduct   385–388

Testing knowledge and understanding

1	a	349	14	a	369–370, 375
2	e	336–339	15	d	370, 371
3	a	341, 343, 346	16	e	372, 396
4	e	349–350	17	e	327, 372
5	b	339	18	c	373–374
6	c	338	19	d	373–374
7	e	356–357	20	c	376–379
8	d	364	21	e	378–379
9	b	363	22	e	384
10	a	364, 366, 371	23	a	386
11	d	369	24	a	386, 390
12	b	375	25	c	391
13	d	370	26	a	391–392

27 e   387, 390, 392, 393
28 d   392
29 e   395
30 b   395

Chapter 10

Testing recall

1 sensory reception, conduction, response   399
2 increases   402
3 brain, spinal cord   402, 404
4 a diffuse nerve net   401
5 two ventral nerve cords, simple sense organs at the anterior end   402
6 double, solid, ventral, many, limited   404
7 dendrites, axons, many, one, sensory neurons, interneurons   401, 404
8 neuroglia, myelin sheath   406–407
9 blood vessels, digestive tract, respiratory system, reproductive system, two, sympathetic, parasympathetic   412
10 somatic reflex arcs, CO_2 414–415
11 accelerated, low O_2, high CO_2 415–416
12 negative, Na^+, into, positively, K^+, out of   418–421
13 sodium-potassium pump, active transport   421
14 faster   407
15 chemical, Ca^{++}, transmitter chemicals, slower   422–423
16 acetylcholine   423
17 reduces, easier, hyperpolarized, IPSP   424
18 integration   425
19 motor end plate, neuromuscular junction, acetylcholine, noradrenalin   426, 427
20 portions of nerve cells, specialized cells, one,   428, 432
21 part of the brain stimulated   430
22 generator potential, action potential, generator potential   429
23 chemoreceptors, specialized receptor cells, neurons   433–436
24 compound eyes, camera-type eyes, molluscs, vertebrates   438–439
25 cornea, iris, reduced   439–440
26 periphery, dim, black and white   440, 441
27 several   441
28 lens, cornea, relaxed, high, thinner   446–447
29 tip, several   451
30 acceleration   451
31 g,b,i,d,f,c,e,h,a   450

32   409, 410

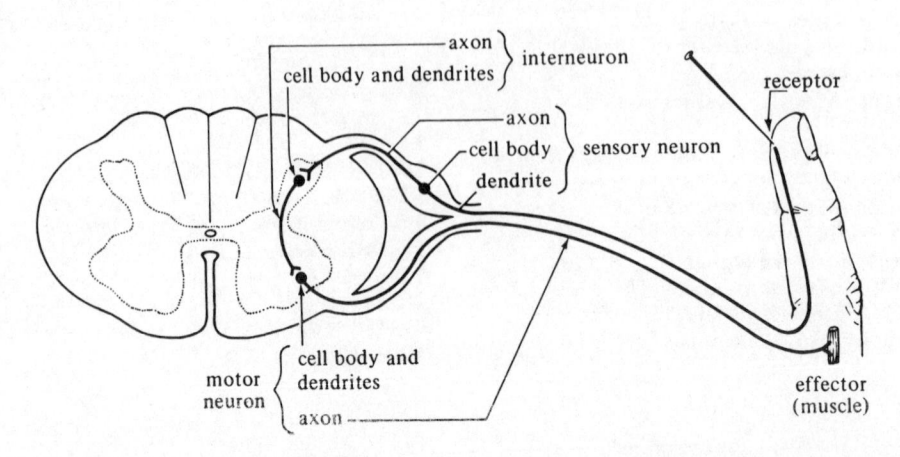

ANSWERS • 227

33	b,c,f,g 454, 457	39	c	457
34	e,g 454, 457	40	b	455, 456
35	a,d,g 454, 457	41	d	454
36	f 455	42	b	455
37	g 456–457	43	c	457
38	a 454	44	b	454

Testing knowledge and understanding

1	c	402	10	d	419
2	b	402	11	d	424
3	b	400, 409	12	c	417
4	a	408–409, 404	13	c	412, 427
5	d	407, 409, 410	14	a	426
6	e	412, 365	15	e	426
7	a	412, 365	16	d	454
8	d	415–416	17	b	455
9	c	418			

Chapter 11

Testing recall

1 470

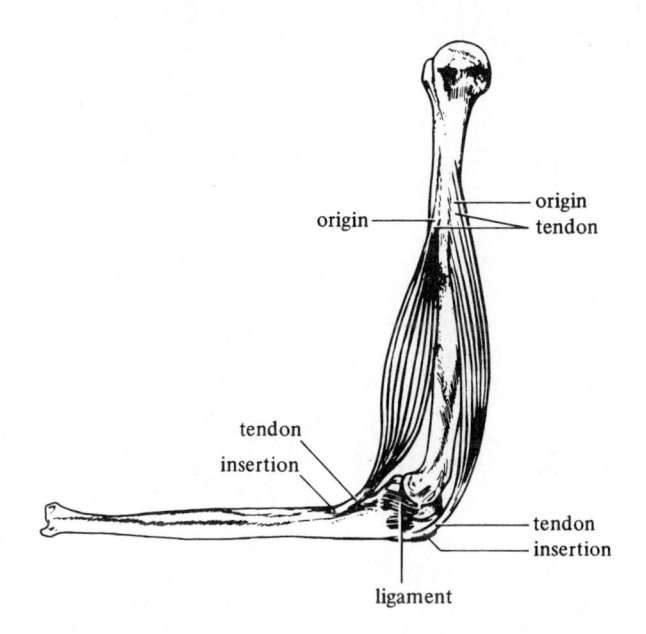

- origin
- origin
- tendon
- tendon
- insertion
- tendon
- insertion
- ligament

2 endoskeleton 468
3 skeletal, cardiac 470, 471
4 skeletal 470
5 smooth, cardiac 470, 471
6 skeletal 470
7 skeletal 470
8 smooth, cardiac 471
9 smooth, cardiac 471
10 smooth 471
11 skeletal 470
12 skeletal 470
13 *a* single twitch: latent period, contraction, relaxation 473
   b summation: muscle does not fully relax during contraction 473
   c tetanus: muscle does not relax at all between contractions; slow, sustained contraction 474
   d fatigue: due to buildup of lactic acid 474

14 The muscle does not fully relax between contractions. 474
15 tetanic contractions 474
16 buildup of lactic acid 474
17 a,b,c,d 474, 475
18 d,a,c,f,e,g,b 482
19 2, a 485
20 2, a 482
21 1, c 486
22 1, b 486
23 2, a 485
24 2, a 485
25 2, a 485

Testing knowledge and understanding

1	c	471	8	c	426
2	e	473–474	9	a	480–481
3	c	474	10	d	481
4	b	474	11	a	475–482
5	c	474–475	12	c	475–478
6	a	474–475	13	e	475–478
7	e	167–174, 475			

Chapter 12

Testing recall

1. Occam's razor 489, 490
2. taxis 491
3. all 493, 500
4. narrower 493
5. habituation 497
6. imprinting 498
7. hypothalamus 501
8. all except happiness 508–509
9. primer pheromones 512
10. threat 516
11. day 522
12. ears 525
13. sun compass, magnetic cues 529, 531
14. generalist 534
15. search image 535
16. home range 537–538
17. both 543
18. polyandry 543
19. do 545
20. displacement 549

Testing knowledge and understanding

1. *e* 501, 503, 506
2. *e* 496
3. *b* 491, 492
4. *e* 498
5. *b* 497
6. *a* 499
7. *c* 497
8. *d* 498
9. *d* 499
10. *a* 492, 497–498
11. *d* 495
12. *d* 498
13. *b* 503
14. *b* 505
15. *d* 506–513
16. *c* 522–523
17. *a* 523
18. *c* 519–521
19. *a* 519–521
20. *b* 519–521
21. *c* 519–521
22. *b* 529
23. *b* 529
24. *b* 529
25. *b* 527, 528, 529, 531
26. *b* 537
27. *c* 549
28. *e* 514, 516
29. *e* 516–517
30. *c* 543
31. *a* 542

For further thought

1. west 529
2. south 529
3. *b* 529

Chapter 13

Testing recall

1. *d* 566
2. *a* 567
3. *c* 566
4. *d* 563
5. *e* 567
6. *b* 560–563
7. *d* 563
8. *b* 562
9. *b* 574
10. *a,b,c* 564, 572
11. *a,c* 564. 572
12. *a,c* 564, 572
13. *a,b,c* 564, 572
14. *b,c* 574–575
15. *c* 578
16. *a* 568
17. *b* 573

Testing knowledge and understanding

1. false—mitosis 578
2. true 574
3. true 562
4. false—not separate; random 572, 581
5. false—twice 572
6. true 578
7. true 570
8. true 569
9. *d* 572, 578
10. *c* 575
11. *e* 575
12. *c* 562
13. *a* 577
14. *c* 572
15. *b* 569–570
16. *c* 581
17. *a* 575

Chapter 14

1. *c* 591
2. *a* 607
3. *b* 584
4. *a* 584
5. *b* 591–592
6. *b* 591–592
7. *c* 593
8. *e* 601
9. *a* 602–603
10. *c* 610–611
11. *e* 610–611
12. *d* 593
13. *b* 584, 610–611
14. *b* 584, 610–611
15. *b* 584, 610–611
16. *c* 614
17. *e* 615
18. *a* 614
19. *c* 615
20. *b* 619
21. *a* 612

Chapter 15

Testing recall

1. *g* 632
2. *c* 627
3. *d* 629
4. *b* 632
5. *h* 630
6. *f* 645
7. *a* 637
8. *e* 633
9. *a* 630
10. *e* 630
11. *c* 630
12. *b* 629
13. *b* 629
14. *c* 630
15. *c* 641
16. *f* 642
17. *g* 646
18. *f* 642

19 a ATGGAAGTCGCA 632
   b AUGGAAGUCGCA 642
   c nucleus 641
   d 4 643
   e UAC, CUU, CAG, CGU 646
   f ribosome 647
   g methionine–glutamic acid–valine–alanine 646

Testing knowledge and understanding

1	b	630–632	9	e	644
2	b	631	10	c	646
3	c	630	11	a	642
4	a	633	12	d	644
5	c	641	13	b	650
6	b	643–644	14	c	619, 652, 664
7	d	632	15	a	58, 646
8	a	642	16	b	651

17 c 101, 641, 644, 652
18 c 664
19 d 658–659

Chapter 16

Testing recall

1	yes	671	15	T, B	689, 690
2	yes	672	16	T	690
3	no	672	17	B	689
4	yes	670	18	B	690
5	no	680	19	C	684
6	yes	677, 679	20	N	684
7	yes	676	21	C	684
8	no	672	22	C	685
9	yes	675	23	N	685
10	T	690	24	C	684
11	B	689	25	N	685
12	B	689	26	C	685
13	B	689	27	C	685
14	T	690			

Testing knowledge and understanding

1	c	669–671	5	b	688–690
2	e	669	6	a	688, 294
3	a	670–672	7	e	689
4	a	676	8	d	692

Chapter 17

Testing recall

1	k	700	8	b	703
2	c	701	9	m	701
3	d	703	10	g	704
4	e	701	11	h	704
5	j	701	12	g	704
6	n	700	13	i	704
7	f	701			

14 a,b,d,e,h,i,k,l,m,o 717–726
15 a,c,e,f,g,h 728, 733–739

Testing knowledge and understanding

1	e	698	10	b	701, 718
2	b	698	11	c	717, 719, 721
3	e	699	12	b	718–719
4	b	698	13	d	718–719, 723
5	c	700	14	d	726
6	b	701, 705	15	c	721, 726
7	e	700–704	16	e	732
8	d	701, 705, 706	17	a	732, 734–736
9	c	714–716			

18 d 675, 728, 735, 744
19 e 744
20 b 744
21 c 745

Chapter 18

Testing recall

1 false—nonrandom 757
2 true 756
3 true 753
4 true 766
5 true 767
6 false—tasteful 773, 777
7 false—to undergo an increase in the frequency of the dark morphs 776
8 true 780
9 false—not change in frequency 756–757
10 true 767
11 true 790
12 true 761
13 false—hybrids may be inviable or sterile 792, 800
14 true 798
15 false—same island   different islands 801
16 false—sometimes 792, 800
17 true 795
18 false—plant than animals 801
19 true 807
20 true 813
21 false—Linnaeus 820

For questions 22–37, page references only are given.

22	760, 768, 788		30	792, 794
23	756, 795		31	797, 795
24	762, 765, 766		32	797
25	778, 780		33	801
26	773, 778		34	813
27	781		35	815
28	783		36	814
29	768, 787		37	813

Testing knowledge and understanding

1	c	754, 758	14	a	795
2	c	756, 757	15	a	792, 808–809
3	a	795	16	d	808–809
4	c	758	17	a	801, 802, 813
5	c	756	18	b	797
6	d	757	19	a	814
7	b	757	20	b	797
8	d	760	21	d	801
9	d	778	22	c	801–802
10	d	771, 772	23	d	813
11	d	781	24	a	819
12	e	787–788	25	e	820
13	a	793	26	c	778, 821

Chapter 19

Testing recall

1	m	839	9	j	827
2	d	822	10	l	822
3	g	833	11	f	840
4	p	855–856	12	c	864–865
5	o	826	13	e	832
6	k	827	14	i	347
7	b	823	15	h	822
8	q	833	16	true	824

17 false—often 826
18 false—r 830
19 false—density-dependent 831
20 true 84
21 false—many 841
22 true 840
23 false—fewer 839
24 true 845
25 false—"Niche" describes the functional role and position of an organism in its ecosystem. 833
26 true 849
27 false—more complex 862–863
28 true 855–856
29 true 864–865
30 false—tundra 869
31 false—more 870
32 true 875
33 false—primary 881
34 true 880
35 false—Palearctic 887–888

Testing knowledge and understanding

2 Young fish died; others had not reproduced yet. 828–829
3 logistic 826
4 K 830
5 inflection point 828
6 density-dependent factors, especially competition 832–833
7 four to five years 826
8 e 825–826

9	e	829	17	a	840
10	b	831–838	18	b	840
11	d	831–838	19	a	840
12	a	839	20	a	840
13	c	840	21	a	839–840
14	d	840	22	c	840–841
15	a	840	23	e	844–846
16	a	840	24	b	849–850

25 e 825, 832, 833, 839, 840
26 b 832–833

27	d	862–863	33	d	865–871
28	c	862–863	34	a	870, 887
29	a	864–865	35	e	877–880
30	d	855	36	d	884–888
31	b	872	37	b	883
32	d	865–871			

Chapter 20

Testing recall

2, 4, 10, 7, 11, 3, 6, 1, 8, 13, 9, 5, 12
894–907

Testing knowledge and understanding

1 c 896
2 a 896, 898, 902, 903, 910
3 e 898

4	c	905	8	e	907–911
5	b	902	9	a	912
6	b	903	10	b	911
7	b	903	11	b	911–915

Chapter 21

Testing recall

1 DNA, RNA 918
2 capsid 917
3 bacteriophages 628, 660, 918
4 nucleic acid or DNA 918
5 ATP 918
6 nucleic acid, protein 918
7 lysis 918
8 whole virion 919
9 RNA transcriptase 919
10 DNA 919
11 reverse transcriptase 919
12 provirus 919
13 lysis 920
14 extrusion 920
15 viroid 922
16 Schizomycetes, Cyanophyta, Prochlorophyta 922
17 procaryotic 922

18 murein 924
19 cell membrane or plasma membrane 924
20 nucleoid 927
21 flagella 926
22 endospores 925
23 ribosomes 120
24 bacilli, cocci, spirilla 923
25 binary fission 927
26 heterotrophic 928
27 obligate anaerobes 928
28 fermentation 928
29 chlorophyll *a* 929
30 Koch's postulates 930
31 toxins 931
32 vaccine 931
33 antibodies 931
34 active 931
35 chlorophyll *a* 934
36 thylakoids 933
37 chloroplasts or chlorophyll *b* 933
38 heterocysts 934
39 binary fission 934
40 Prochlorophyta 936

Testing knowledge and understanding

1	*b*	920	9 *d*	129, 922
2	*b*	920	10 *a*	189, 929
3	*b*	917–920	11 *c*	925, 927, 929
4	*e*	918, 920, 921	12 *c*	932
5	*e*	919	13 *c*	921, 930–931
6	*a*	921	14 *c*	934
7	*b*	917, 918	15 *b*	934
8	*d*	120, 922	16 *b*	934

Chapter 22

Testing recall

Chart:
 Mastigophora 940–941
 Sarcodina 941, 943
 Sporozoa 943–945
 Ciliata 945–947
 Protomycota 948
 Gymnomycota 948–951
 Euglenophyta 952–953
 Crysophyta 953–954
 Pyrrophyta 955

Testing knowledge and understanding

1 *d*	943–945	8 *a*	948–951
2 *c*	939	9 *b*	939, 948
3 *e*	943–945	10 *a*	953
4 *a*	944	11 *a*	953–955
5 *c*	945–947	12 *c*	953
6 *c*	943–945	13 *e*	955
7 *c*	941, 949		

Chapter 23

Testing recall

1 *a*	957	6 *b*	968
2 *b*	965	7 *c*	968
3 *a*	957	8 *a*	957
4 *c*	968	9 *c*	968
5 *b*	965–968	10 *a*	957

11 E (and a little in T–the kelps) 956
12 T 959
13 E (and occasionally in T–some brown algae) 970
14 E 956
15 T and E 970   17 E 970
16 T 956   18 E 970

Chart:
 Bryophyta 971–972
 Psilopsida 974–975
 Lycopsida 975–977
 Sphenopsida 977
 Pteropsida 979–980
 Spermopsida 980

Testing knowledge and understanding

1 *c*	959, 960	13 *c*	974–975
2 *d*	962	14 *e*	981–986
3 *e*	965–968	15 *a*	986–987
4 *a*	952, 971–972	16 *b*	988–989
5 *d*	968	17 *a*	989
6 *d*	965–968	18 *c*	988
7 *a*	957	19 *a*	987
8 *b*	956, 968	20 *b*	988
9 *c*	974	21 *e*	987
10 *a*	984	22 *c*	988
11 *d*	989	23 *c*	988, 989
12 *c*	956, 971		

Chapter 24

Testing recall

Chart:
 Oomycota 992
 Zygomycota 992–993
 Ascomycota 994, 997
 Basidiomycota 998

Testing knowledge and understanding

1 *a*	992	5 *c*	998
2 *c*	992	6 *e*	998
3 *a*	993	7 *d*	998
4 *c*	993		

Chapter 25

Testing recall

1 Coelenterata, Ctenophora, Echinodermata 1003, 1038
2 Echinodermata   1038–1039
3 Arthropoda   1029
4 Brachiopoda, Phoronida, Ectoprocta   1021
5 Chordata   1045
6 Mollusca   1022
7 Aschelminthes, Acanthocephala, Entoprocta   1018
8 Echinodermata, Hemichordata   1043
9 Platyhelminthes   1011
10 Porifera   1000–1001
11 *c*   1058
12 *a*   1047
13 *k*   1045–1046
14 *h*   1059
15 *g*   1042
16 *j*   1052
17 *i* (also some Amphibia)   1050
18 *d*   1046
19 *h*   1059
20 *e*   1050

Testing knowledge and understanding

1 *a*   1001
2 *b*   1003
3 *c*   1015–1018, 1042
4 *d*   1022
5 *d*   1022   9 *c*   1029–1030
6 *b*   1019   10 *d*   1005, 1043
7 *c*   1003   11 *e*   1020
8 *e*   1011   12 *d*   1052
13 *d*   1029–1030, 1053, 1058–1059
14 *c*   1022, 1044
15 *e*   1043   18 *c*   1047–1049
16 *e*   1043, 1045   19 *c*   1051
17 *e*   1050, 1052   20 *c*   383, 1053
21 *c*   1053
22 *a*   1059
23 *e*   1058   27 *a*   1066, 1067
24 *d*   1051   28 *c*   1065
25 *d*   1060   29 *e*   1061
26 *b*   1069